Tractionology

WANT MORE AND BETTER CLIENTS?

READ THIS BOOK

JOE STOLTE

TRACTIONOLOGY:
WANT MORE (AND BETTER) CLIENTS? READ THIS BOOK

ISBN 978-1-7365575-0-1

Dedication

To Judy and Journey: Thanks for letting me hide away in my "office" to write this book. You're my world, my 'why', my everything. I love you both. Here's to making our future bigger than our past, one day at a time.

To Chris Smith: Thank you for seeing potential in me when I couldn't see it in myself, over and over again. Thank you for showing me how to be a professional while still being cool at the same time.

To Eben Pagan: Thank you for introducing me to meditation, NLP, and direct response marketing. From one Oregon farm town kid to another, I appreciate you.

To Frank Kern: Dr. Kern, thank you for teaching me about the power of generating goodwill in a market and for making it all look so easy. Oh and thanks for letting me shamelessly borrow your ad headline for the title of this book. You're a legend #NASA.

Table of Contents

Who This Book Is For

To the bridge builders. The helpers of other humans. The consultants, the coaches, the experts, and the professional service providers. The people who are in the business of helping others get from where they are to where they want to be. This book is for you.

In case you missed the title, this book has one very clear objective: to show you exactly how to get more (and better) clients.

The kind of clients that show up to your strategy sessions and sales calls and say things like "I've been looking for someone just like you", "I already know I want to work with you", and "how do we start working together right now?".

When was the last time you heard a prospective client say something like this to you?

The sad reality for most client-based businesses is that it's far more common to hear things like "I need to think about it" or "it's too expensive" or "I need to talk to my spouse or business partner".

Or even worse, you have very few leads and very inconsistent sales calls and strategy sessions on your calendar.

If this sounds familiar, I have good news for you.

By the time you're done with this book, you will know exactly how to create a consistent stream of high-paying clients coming to you, asking to work with you.

Clients who are ready to pay up-front and who are prepared to do the work.

I'm going to take you step by step through the mental models and frameworks I've used over and over again to help my own clients close more than $10M in sales during the 2020 pandemic and economic chaos.

You're going to learn exactly what to do and the exact order to do it depending on what stage you're at in your service-based business.

But first some caveats.

This book is not for someone who is just getting started and doesn't have a real business and paying clients.

This book is articulated for the consultant or coach or freelancer or service professional who has clients but has gotten stuck on the six-figure hamster-wheel and wants to break through to $100K+ in revenue per month.

If you're a beginner with no clients, you will get value from reading this book. Just know in advance, I didn't have you in mind when writing it.

In fact, I very specifically used the title of this book to call out to people who already have clients (a trick I'll teach you later in the book).

If you have a real business with real clients and the courage to take action on what you learn from this book, what I have to say could change your life.

This material works for coaches, consultants, freelancers, insurance agents, mortgage brokers, wealth managers, attorneys, and anyone who is in the 'client business' or professional services world.

But here's the catch.

Your results will vary based upon your courage to take action, your patience to validate things with data, and your work ethic.

Running a service-based business requires taking risks, putting in consistent effort, and taking massive action.

I have stripped things down to the essential, gotten rid of the hype, and made the formula as process-driven and scientific as possible. If you're not willing to put the work in though, you will not get the results.

With that said, let's jump in and start the show.

Introduction

In case we've never met, let me hurry up and get my mercifully short, self-aggrandizing introduction out of the way.

But first, let me tell you who I'm not.

I'm not a speaker, I'm not a 'best-selling author', I'm not a guru, and I don't have a software platform to sell you on after you're done reading this book.

In fact, I don't own a Lambo, I don't own a private jet, I don't have a real estate fund for you to invest in, this isn't a get rich quick scheme, and I don't have an MBA or any other letters before or after my name or any other fancy credentials.

I do have a pretty good track record of being a consultant and an entrepreneur though.

In the last 10 years, I've generated over $1 billion in enterprise value for my clients and through my own companies.

I've done consulting projects for Dara Khosrowshahi (now CEO of Uber), Satya Nadella (now CEO of Microsoft), and even the Head of Defense and Space at the largest airplane manufacturer in the world.

I've also built and scaled two venture-backed software companies to 70+ employees and hundreds of millions of dollars in value.

In 2020 I helped about a dozen small business clients generate over $10M in sales while wrestling with panic, pandemics, lockdowns, and economic free fall. All while freeing up their days so they can spend more time helping their family and their community.

Today I run a growth-consulting company called The Tractionology Group. We help purpose-driven consultants, coaches, and service professionals get more (and better) clients without complicated advertising or confusing technology.

And perhaps most importantly. I know that absolutely none of what I just said really matters. I fully understand and appreciate that you are here for you and to grow your client-based business.

This book is not about me, it's about helping you.

So, without further ado, let's do exactly that, shall we?

SECTION I

GETTING STARTED FROM THE INSIDE OUT

Chapter 1

THE SCIENCE OF BELIEF

Somewhere between blowing it with my first technology start-up and going on to launch a marketing agency that I eventually sold for a healthy profit, I started studying neuro-linguistic-programming (NLP).

One of the big mental models I took from NLP was the concept of 'modeling', which involves identifying someone who is successful, deconstructing their success into smaller parts, and mimicking these smaller parts for yourself to replicate the success of the subject that you're modeling.

One of these 'smaller parts' that are helpful to understand when modeling someone else is their beliefs and values.

Nail these and you're on your way to successfully replicate the excellence of any top performer.

For example, a (virtual) mentor of mine, Wyatt Woodsmall, did a modeling study with the U.S Military called "Project Jedi" to help soldiers improve their accuracy in shooting .45 caliber pistols.

Wyatt and his team went in and modeled the army's top experts in marksmanship, deconstructed their behavior, and most importantly what they believed.

At the end of the study, they were able to take a female soldier who had never fired a pistol before and have her replicate the results of the expert marksman in only 8 hours.

They were transforming unqualified trainees into experts in half the time with less ammunition spent in the process.

One of the core beliefs they found in the expert marksman vs. the nonexperts was that the experts believed that guns are safe and that it is natural and safe to learn how to shoot.

This was in direct contrast to the non-experts who believed that firearms are bad and are used to kill innocent people.

By installing the right beliefs and mindsets into the nonexperts, they were able to get radically improved results in less time.

You might be thinking, Joe, what the heck does this have to do with helping me get more (and better) clients for my business?

Whether you believe you can or believe you cannot you are right. Beliefs can be either supportive or disabling.

Beliefs are a critical factor in marksmanship as in any other discipline, like learning how to get more (and better) clients.

If you don't have the right beliefs, the strategy and mechanics I'm going to show you will be absolutely worthless.

In fact, they might be worse than worthless, they might actually cause more harm than good as you torture yourself into trying to apply them only to be destined for failure.

Before we jump into the core framework in this book, you need to understand the 5 mindsets and beliefs of getting more (and better) clients.

Chapter 2

THE 5 MENTAL MODELS YOU NEED TO WIN

#1. MAKE EVERYTHING ABOUT THE CLIENT AND WHAT THEY WANT.

Zero prospective clients want your coaching, consulting, or services. They want the results your services are going to deliver them. Your 'shiny stuff' is actually friction in the way of getting what they want.

In fact, people only become clients because they believe you can help them get what they want.

In all of your marketing materials, sales conversations, copy, social media posts, emails, and communications make EVERYTHING about the client.

Specifically, make everything about what your client wants and how they can get it.

Clients don't care about you. They don't care about your services. They don't care about your Lambo or jet photos or your pretty social media profile.

Case in point, look at what got you to read this book. I didn't call this book "The wonderful adventures of Joe Stolte and his fancy business program".

Assuming you're not a family member or friend who I bribed to read this book and give me feedback, you're probably reading it because you want to scale your services business by getting more (and better) clients.

If you only took one thing from this book and went off to get more (and better) clients, make it this mindset shift. Everything is about the client, what they want, and how they get it. It's not about you or your services or your products.

#2. THE BEST WAY TO DEMONSTRATE I CAN HELP CLIENTS IS BY ACTUALLY HELPING THEM.

One of my favorite marketers from is a man named Eben Pagan. Eben came up with this idea called "Moving the Free Line".

This was a contrarian idea that flew in the face of the common thinking at the time, which was to "charge for everything and save your best stuff for your higher ticket coaching or consulting programs".

Eben had the genius idea to go find your best content and give it away for free upfront to dramatically reduce the amount of time it took to get an ice-cold prospect to trust you enough to buy from you.

At the heart of what Eben was teaching with 'moving the free line' was the idea that the best way to demonstrate you can help clients is by actually helping them.

One of Eben's peers, the legendary Frank Kern, expanded on this concept and called it "results in advance" and made it the core thesis of his marketing strategy.

Billions of dollars in enterprise value have been created in silicon valley and within the software-world using this concept as a now common business model called "freemium".

Give your best stuff away, demonstrate value, shorten the trust timeline, convert more clients by actually helping them.

Don't hide your best 'tricks', use them in your marketing and sales process to help your clients. Otherwise, one of your competitors is going to start doing this and leave you exposed.

#3. THERE ARE NO SHORTCUTS. I NEED TO DO THE RIGHT THINGS AND LET THE RESULTS CATCH UP.

There is no magic pill. There is no easy button. You are not "one funnel away" or one sales letter away, or one anything away from business success.

Success is an emergent property that comes from doing multiple things right at the same time, not from doing one simple hack or shortcut.

It's kind of like 'health' when it comes to your body. Your body is made up of a bunch of organs doing their individual jobs at the same time. Below the organs you have tissues, below the tissues, you have the individual cells, and so on.

Health is what happens when all these layers and systems are working at the same time. You can't shortcut good health. The same is true in your business.

For you to get more (and better) clients and to scale your client-based business, you need to learn how to do many things right at the same time. You need to be patient as you put these things in place because no single tactic or strategy is going to produce life-changing results overnight.

If you go to the gym 5 days in a row and stop, you don't magically transform your flabby stomach into a cover model-level physique.

If you plant seeds in your garden, add some water, and leave them overnight, you don't get an apple tree that bears fruit the next day.

We know this to be fact, and yet we keep fooling ourselves into thinking there's a magic pill that will grow our business overnight.

The sooner you make the mindset shift away from "push-button-results" to "do the right things and the results will catch up" the sooner you will be on the path of getting more (and better) clients.

#4. I'M 100% RESPONSIBLE NOW AND NO ONE IS COMING TO SAVE ME.

If you want to win as an entrepreneur, you need to take full responsibility.

It took me a long time to learn this lesson but when I finally started to take responsibility, I got results a lot faster.

Once upon a time, I used to produce really big dance events. A few years ago, I was producing one of the biggest break-dancing events in the world, the North American Continental Qualifier for the Youth Olympic Games.

Hundreds of staff, thousands of dancers from all over the world, more than $1M in production on the line, and I was the man in charge of it all.

Things did NOT go perfectly.

The elevated stage we created started to collapse in the middle of the event (my fault).

The judges responsible for evaluating and ranking the dancers forgot to submit the top-ranked dance crew with their final results (my fault).

The biggest sponsor was not happy with the lighting and asked me about 10 times to make adjustments all throughout the show (my fault).

And on and on.

As the leader, I took 100% responsibility for myself, the event, the staff, the production, everything.

When things broke or didn't work or didn't go as planned, I pointed the finger in one direction: at me.

Full, 100% ownership. And guess what? Everything worked out. Because I was focused on understanding the issues and finding solutions.

Not on blaming others or waiting for my guru to give me the answers.

This is exactly how you need to run your services business. Take ownership and accountability for everything.

Prospects don't have money? Your fault.

Don't have enough qualified leads? Your fault.

Clients not paying on time? Your fault.

The sooner you take full responsibility, the sooner you will learn the lessons you need to level up and get more and better clients to scale your business.

#5. FOCUS, FOCUS, FOCUS. MAKE IT WORK SMALL AND THEN GO BIG.

As entrepreneurs, we love to skip steps and chase shiny objects. By nature, we believe the rules don't apply to us. In most of these cases, we are wrong.

That's why the business failure rate is astronomical.

Having a successful client-based business is about cracking the code and validating things one piece at a time.

90% of my clients, before we start working together, have too many offers, no clarity on their ideal dream prospect, and are completely distracted trying to do 50 things at once.

My hope for you after reading this book is that you can see how radical simplification is best path to achieving business success.

We'll cover this in detail in the rest of the book but it's worth mentioning here.

If you aren't consistently generating $100K a month or more in revenue yet, you probably need to go back to the basics and focus on 3 simple things:

1. Getting more qualified prospects on sales calls

2. Converting those prospects into clients for your best core offer

3. Delivering results with world-class fulfillment of your offer

Anything else you are working on that doesn't directly relate to or support these 3 things is robbing you of your energy.

Once you have these basics down, you can scale to the moon.

As Peter Drucker once famously said in his book, The Effective Executive, "*First things first, second things not at all*".

This book will show you the first things you need to master to keep you focused so you can get more (and better) clients to scale even faster.

We covered a lot of ground in this chapter. Here's a brief recap of the mental models and beliefs so they're fresh in your mind:

1. Make everything about the client and what they want.

2. The best way to demonstrate I can help clients is by actually helping them.

3. There are no shortcuts. I need to do the right things and let the results catch up.

4. I'm 100% responsible now and no one is coming to save me.

5. Focus, focus, focus. Make it work small and then go big.

Chapter 3

TRACTIONOLOGY: HOW TO GET MORE (AND BETTER CLIENTS)

Getting more (and better) clients is not complicated and it doesn't have to be hard.

If you read the last chapter and have committed to adopting the 5 mental models and beliefs you need to win, you're well on your way. The rest of what you need is all mechanics.

I call these mechanics "Tractionology", which is a 4-step framework for breaking down your client-based business into its component parts and rebuilding each part to consistently get you more (and better) clients.

Each step in the framework builds on the last step and if you're having trouble getting more (and better) clients, you've probably skipped a step or have not built a solid foundation.

Step 1 - Package. Before you can turn your business into a high-quality client transformation machine, we need to examine the very core and foundation of your business. This foundation includes narrowing the focus of your client avatar (who you sell to), identifying your best core offer (what you sell), and crafting your core content (what you say). We need to make sure these are dialed-in and packaged correctly before we do anything else.

Step 2 - Attract. Once you have your packaging ready, we need to pick the right client attraction channel based on where you are in your business. The focus here is to have you focus on the right channels, one by one, instead of doing 10 things at once and getting overwhelmed. Think of this as a menu that you can pull from to generate more qualified leads based on where you are in the evolution of your client-based business.

Step 3 - Convert. With your foundational packaging in place, and your client attraction channels working, you should be getting high-quality prospective clients showing up for sales calls. In this section, we're going to dive into how to set up the perfect sales call that enrolls clients quickly and has them excited to pay you up-front and get started working with you right away.

Step 4 - Transform. Once you have high-paying ideal target prospects converting into clients, your job is just getting started. In this phase, you'll learn how to script the initial interaction with your clients in such a way that gets them transformative results and turns them into raving fans who can't wait to send you a mountain of qualified referrals.

Like all good frameworks, the 4-step Tractionology process forms an easy-to-remember mnemonic. This mnemonic spells the word P.A.C.T.

When you look up the definition of the word 'pact', our friends from the Merriam-Webster dictionary define it as "a formal agreement between individuals or parties". Which is fitting given that in order to get more (and better) clients, you need to form a pact with your ideal target client in order for them to do business with you.

Tractionology is the science of getting more (and better) clients for your service-based business. Unlike traditional marketing and sales methods that rely on getting lucky with great creative or using pushy sales techniques, Tractionology shows you how to validate exactly what works in your market to get more (and better) clients.

Here's my goal for you with this framework. I want to teach you how to think. If something isn't working for you, you can use this framework to go back and examine where you're falling short, make adjustments, validate those adjustments with the market, and iterate until you crack the code on what works.

There are layers to this game. What gets you to $30K a month won't get you to $100K a month. What gets you to $100K a month won't get you to $100K a week. If you follow the steps, try new things, and validate your results within the laboratory of your market, there's no limit to how far you can scale your service-based business.

SECTION II

BUILDING YOUR ROCK SOLID FOUNDATION

Chapter 4

A COUNTERINTUITIVE PLACE TO START

My wife and I love to travel. Before we had our son, we spent about 150 days a year on the road traveling to other cities and countries and exploring other cultures. In 2013 we went to Malaysia for the first time and stopped by the Petronas Towers in Kuala Lumpur.

The Petronas Towers are two beautifully architected massive towers that stand 88 floors high. In fact, from 1998 to 2004 they were the tallest buildings in the world.

What's interesting to me about the Petronas Towers isn't just that they are 88 floors high, it's that their foundation is 36 stories deep. About 26% of the total above-ground height was dug out below the building underground in order for it to stand firmly as the tallest set of buildings in the world.

It's counterintuitive that in order to build the tallest building in the world, you start by digging a 400-foot hole in the ground.

That's sort of what it feels like to have an established business and to have your coach ask you to go backward and redefine your market, your offer, and your messaging.

It's counterintuitive to go back to the beginning and check the foundation of your business before we start giving you strategies to get more clients.

And just like building the tallest building in the world, getting the foundation of your business rock solid is critical if you want to get more and better clients.

The foundation of your business starts by understanding your market. Also known as your ideal target client. Put simply, we need to double-click on WHO you sell to and make sure you understand them and their evolving needs thoroughly.

Most of the time when I ask a client who their ideal target client is, they immediately launch into all the different services they offer and all the different target markets they serve, which is a huge red flag.

There are endless "consultants" who offer book funnels, events, one on one coaching, digital courses, and complex "value ladders" who have not figured out how to crack $100K a month consistently.

"Full service" agencies that will run your paid ads, design your brand, and shine your shoes.

Insurance agents who sell every kind of policy to literally any human with money who can pass a physical exam.

Wealth managers who will hunt down the IRAs and investable assets of all humans with a college degree and a steady job.

If it feels like too much, it's because it is too much.

In case you haven't noticed there's a war for human attention being fought by social media algorithms, content creators, mobile phone manufacturers, and hungry marketers eager to direct your focus towards their new shiny thing.

You can't win with broad. You have to go narrow.

In fact, to make it crystal clear, if you aren't yet consistently producing $100K a month in your client-based business and you have more than one offer and more than one ideal target client, there's a 98% chance you are:

- Weighing your business down with too much complexity

- Not focused and lack a clear marketing message

- Competing on price with a lot of competitors (and losing deals)

There is a time in the trajectory of your business for having more than one offer and serving an adjacent complementary market, but if you're not currently getting more of the target clients you really want then you need to focus on building a solid foundation. This starts with getting clear on who your ideal target clients are.

Chapter 5

IDENTIFYING YOUR IDEAL TARGET CLIENT

I have two powerful questions that, if answered correctly, will immediately help you simplify your business to the essential and strip away the distractions that are diluting your focus, robbing you of energy, and making you less potent.

These questions, if you spend time with them and go deep, will transform your marketing and sales from an unfocused flood-light into a laser beam of ethical influence.

The Ideal Client Focus Question #1: If you could only get paid AFTER your client got the results you promised, what relevant characteristics would they have to have in order for you to take them on?

This question is powerful because it forces you to think about your best clients, the ones that show up on time, do the work and consume your guidance, and get the best results.

Let's use my business, The Tractionology Group, as an example. We help purpose-driven consultants, coaches, and service professionals get more (and better) clients without complicated advertising or confusing technology.

About a year ago, I sat down and answered the Ideal Client Focus Question to better understand who our ideal target client is and determined they have 3 core characteristics:

1. **They have an established business that serves clients.** We're a terrible fit for beginners or people who are just getting started. We're a slam dunk if you serve "clients" and have some success in your business already. This is important because we want to sell to people who have money, not people who need to go into debt to work with us. It's also important because beginners have entirely different challenges that require radically different solutions.

2. **They have some level of goodwill in their market**. Most of the clients that come to us have gotten results from working with their own clients in the past. This is a good sign that they are far enough on their trajectory and will be a good fit for us. This trait also serves as a proxy indicator that they have the right raw material or assets in their business that we can work with to help them grow.

3. **They take fast action**: 100% of our best clients take fast action. They don't overanalyze things, they don't wait for the perfect conditions, they have a track record of getting an idea and running with it. They are 'doers' and love to get stuff done.

The Ideal Client Focus Question #2: What are the defining mindsets and values that your best clients share with you?

This question is a bit different and requires you to understand the mindset of your ideal client.

Let's use my company again as an example to illustrate.

We work really well with and can consistently generate results for clients who have the following mindsets:

- **Purpose Driven**: People who aren't focused on getting rich quickly. Yes, they want to have financial abundance but they also want to make an impact on their market and they actually care about helping other humans and making an impact in the world. These folks play the long game and are willing to do the right things and let the results catch up.

- **Growth Mindset**: Our best clients believe in hard work and their own capacity to grow by way of their own effort. They have an abundance mentality and aren't scared to try new things and they don't get paralyzed by fear of failure. This is critical otherwise they won't act on our recommendations and won't get results.

- **Love Sales**: They don't need to be world-class at selling but they need to be warm to the idea of sales and marketing. We are not a good fit for the 'woo' crowd that thinks selling is evil and companies are somehow inherently bad. We love these people, they just don't make great clients for us.

It should take you about 60 minutes to come up with this list for your company. I recommend you go back and analyze all the clients you've done business with and look for clues around why some clients bought without friction and got results and why others did not. I've taken dozens of clients through this exercise and every single one of them has found patterns that have been "hiding in plain sight" waiting for them to be uncovered.

IDEAL CLIENT FOCUS EXAMPLE

Earlier this year I had the opportunity to coach a client in the wealth management space. This is a market that's highly commoditized and the majority of the companies are competing for the same clients with the same broad and aimless messaging.

This client, we'll call him "Barry", was different. He got laser-focused on his market and came up with the following characteristics:

- People of color (who Barry was called to serve)

- Graduating residency and going into full-time work as physicians (e.g. they have money)

- Don't know how to manage their money

- Have high student debt

Barry's ideal target client's core mindsets and values are:

- Growth mindset

- Care deeply about their financial future

- Driven to lead (they were usually the leader of their residency for example)

With this level of clarity serving as the foundation, I was able to work with Barry on deploying just one new powerful attraction channel that ended up driving a 52% increase in his revenue. There was a bit more involved in the process (which you will learn about later in this book) but the foundation of this increase was born out of getting hyper clear on defining his ideal target client.

Take a few minutes and write down your initial answer to the two ideal client focus questions (first thought best thought, just write what comes up for you).

- **The Ideal Client Focus Question #1**: If you could only get paid AFTER your client got the results you promised, what relevant characteristics would they have to have in order for you to take them on?

- **The Ideal Client Focus Question #2**: What are the defining mindsets and values that your best clients share with you?

With these answers in mind, let's dive into the next step of the "packaging" phase of your journey through the Tractionology framework, where I'll help you identify your best offer for your newly focused ideal target prospect.

Chapter 6

YOUR CORE OFFER ANALYSIS

Wouldn't it be cool if you could figure out the one core offer (service or product) you can sell your ideal dream client that:

- Pays you the most money

- Energizes you

- Is easy to deliver

- And that you actually enjoy doing?

Great, because that's exactly what we're going to do next.

Our mission in this section is to help you analyze your offer, strip away the distractions, and boil it down so it meets some very specific criteria designed to keep you focused, happy, and making money from more (and better) clients.

We're going to get you away from chasing shiny objects that rob you of all your energy and help you get laser-focused on your best core offer.

A word of caution before we proceed here though. Your mind, your team, your ego, maybe even your spouse will tell you're crazy for cutting off revenue-producing parts of your business.

You will need some real courage to say no to the wrong clients who are willing to pay for the wrong offers so we can get you focused on the right path.

Remember, in the "Package phase" we are taking you from a floodlight to a laser beam. Less is more. Stay focused, trust the process, and let's get started.

I'll go through each of the core criteria, then I'll show you a simple worksheet you can use to do this for your business.

Criteria 1: Highest net profit. This one is pretty simple. What do you currently sell that pays you the highest net profit? You might have to pull up your profit and loss statement or do some 'back of the envelope math' to figure this out for each product or service you offer.

I fully realize that by asking you to do math I may lose about 80% of the people reading this but I promise this is easy (and fun).

Now the first thing your brain is going to do when you go through this exercise is to say, "well whatever pays me the most money, that's my core offer" and I hate to say it, but your brain is wrong.

I've fallen into that trap before as have a lot of my clients. The goal here is not only to get paid but to get energy from delivering your services and to enjoy life while doing it. Otherwise, you'll end up turning your business into a high-paying job that you hate.

Criteria 2: Energizing. Which offer do you bring to the market that brings you the most energy? Do you have something that you offer that really gets you charged up when doing it?

A big part of success is being happy in life. If you're working on stuff that doesn't energize you, you're not going to be able to play the long game. You're going to burn out and give up.

For example, I really enjoy 1:1 coaching but I LOVE teaching in front of groups. I could teach and coach in a group setting for hours and hours and still feel energized afterward. I'm an extrovert who loves connecting with and lighting up other human beings. This is totally subjective and personal and as long as you're honest, there are no wrong answers here.

Criteria 3: Easy to Deliver. How easy is each service that you offer to deliver? For example, I like writing copy and coming up with ad angles but it's really hard for me to do either of these activities quickly. It takes me forever. Having clients show up for a private day with me where we use the Tractionology framework to get their business to $100K+ a month, that is almost effortless for me.

Criteria 4: Easy to Sell. How much work do you have to do to sell this offer? Is it easy to get prospects to say yes or do you have to throw in 50 bonuses and twist their arm to get them to buy?

At the time of writing this book, there's a global pandemic keeping people locked in their homes. Getting people to pop over to my house from around the world for a private day is pretty close to impossible.

That being said, with so many people isolated and starved for interaction, getting them to sign-up for a 1:1 or group coaching program where they get to associate with other powerful entrepreneurs and get direct accountability is pretty frictionless.

Criteria 5: Future Scalability. Once validated, can you easily automate or delegate the delivery of this offer?

The key words here are "once validated", meaning you have cracked the code on the rest of the Tractionology framework and clients are lining up around the virtual block to buy your offer.

Going back to my business as an example, I can easily train other coaches to run a 1:1 program or turn that program into a training course to scale in the future. Having people come over for private days to work with me personally on the other hand is far harder to scale because it requires me to deliver it.

Now that you understand the criteria, the next part is really simple, and I've created a worksheet to help you bring it all together.

You can get access to this worksheet by going to tractionologybook.com/resources.

Once you have access to the worksheet, simply list each of your offers in the left column, and then rank each one with a score of 1-10 for each of the 5 criteria and add up the totals.

When you're done ranking each offer and adding the totals, look at the final scores to see which one ranked the highest. Congratulations, this is your best core offer!

Offers	Criteria 1 Profit		Criteria 2 Energy		Criteria 3 Ease (Delivery)		Criteria 4 Ease (Sale)		Criteria 5 Automate / Delegate		Total -
Virtual Group Coaching	8	˅	9	˅	8	˅	8	˅	6	˅	39
1:1 Coaching	7	˅	8	˅	7	˅	10	˅	6	˅	38
Digital Courses	8	˅	6	˅	8	˅	5	˅	10	˅	37
Books & digital downloads	6	˅	5	˅	10	˅	4	˅	8	˅	33
Private Client Days	7	˅	10	˅	10	˅	1	˅	2	˅	30
"Done For You" Consulting	3	˅	1	˅	1	˅	10	˅	1	˅	16

Example: Completed core offer analysis for a coaching practice

Now that you have your ideal target client and your core offer in hand, it's time to craft your minimum viable magnetic messaging.

Chapter 7

CRAFTING YOUR MINIMUM VIABLE MAGNETIC MESSAGING

When I left Microsoft as a full-time employee and I wandered into the world of Silicon Valley and technology start-ups, one of the first things I noticed was that all my new peers spoke a different language. I started hearing words like unicorns, pivot, freemium, product-market-fit, CaC, LTV, and so on.

Having worked in tech, I understood most of these words and concepts, but I had to play catch-up on some of the methodologies where they originated from as I built my first software company.

Some of these methodologies included agile, customer development, and the lean start-up framework. At the core of these paradigms was a scientific approach to rapidly hypothesize, validate, iterate, and scale a software or product-based business.

At the same time, my internet marketing and eCommerce friends were following in the footsteps of brilliant Madison Avenue direct-response marketers such as David Ogilvy, Claude Hopkins, and Eugene Schwartz, who also happened to follow a scientific approach focused on marketing and messaging.

At the core of what my tech friends and what my internet marketing friends were doing were coming up with ideas, testing those ideas with real customers, and iterating based on the data they got back. One camp was doing it with software products and apps, the other was doing it with ads, emails, social media posts, and snail-mail.

What does this have to do with crafting messaging for your service-based business? Everything my friend.

At the core of what the guys in Silicon Valley and the Direct Response folks on Madison Avenue were doing was "making it work small before they went big" (see mental model #5).

They were getting feedback from the market, forming a hypothesis, validating with real customers and clients, iterating quickly, dispassionately discarding what didn't work, and doubling down on what did work.

That is exactly what you'll need to do in order to craft specific, focused, razor-sharp messaging so you can get more (and better) clients.

The tool you will use to start your journey of crafting the perfect message for your market is what I call your Minimum Viable Magnetic Messaging:

- **Minimum Viable**: The minimum required to get the intended objective.

- **Magnetic**: It resonates perfectly with your ideal target client and attracts them to you.

- **Messaging**: The actual words and content you will deliver to your market

Your minimum viable magnetic messaging doesn't need to start out anywhere near perfect. Like my software and direct response friends, we're going to use a simple process of validation and continuous improvement to upgrade your message over time.

To do this we're going to do the following:

- **Hypothesize**: Come up with a hypothesis around the best messaging and content for your market using 2 simple frameworks.

- **Validate**: Validate the messages by talking to actual customers.

- **Iterate**: Keep track of what works, what doesn't work, and what trends you are seeing so you can keep updating your messaging until it's razor-sharp.

With that said let's take an educated guess as to the most basic of messages, you can use to get the attention of your target prospects long enough to sign-up for a sales call or strategy session.

MESSAGING FRAMEWORK #1: YOUR OFFER POSITIONING STATEMENT.

I help (target prospects) get (desired result) by (your offer or secret sauce).

If this framework looks familiar, it might be because I've been using it to describe my consulting company all throughout the book.

"We help purpose-driven consultants, coaches, and service professionals get more (and better) clients through a proven framework called Tractionology"

If you don't (yet) have a distinct framework for how your services-business gets client results, you can also use an alternate version of the offer positioning statement.

I help (target prospects) get (desired result) without (the painful thing they want to avoid).

In the case of our company, here's what that looks like.

We help purpose-driven consultants, coaches, and service professionals get more (and better) clients without complicated advertising or confusing technology.

My friend Caleb runs an advertising agency called "Patient Autopilot" and his basic positioning statement is "We help chiropractors get and keep patients using performance-based marketing".

One of my former clients, GrowFlow (a subscription software company) has the following positioning statement "We help licensed cannabis operators manage and grow their business with world-class software and support".

Short, simple, effective.

When I ask most business owners what their business does, I usually get a 5-minute, incoherent barf of words that requires multiple follow-up questions to really understand what they are trying to say.

They launch into their fancy method or product or process or their secret sauce and ignore the market, the result they want, and how they help them get it.

To make sure you don't fall into this trap, leverage either the primary or the alternate framework provided above to help you get crystal clear and concise on your offer positioning statement. You can do this for your core offer or for your business as a whole.

Don't worry about making it perfect, you're going to validate this by talking to actual customers later.

MESSAGING FRAMEWORK #2: YOUR CLIENT VALUE MAP

The objective of the Client Value Map is to identify the best content and messaging you can deploy into the marketplace to attract your ideal target clients. This content and messaging are the mechanisms you will use to demonstrate to your ideal target clients that you can actually help them by solving some of their biggest problems in advance.

The thinking behind this framework is that the more you move your ideal target clients towards the actual results they want before you ask for the sale, the easier it will be to enroll them as clients because you've already demonstrated massive value.

To craft your Client Value Map, you simply need to answer this question:

If you were working with your ideal target client, what are the three to five big things you would do to get them the fastest and most significant results?

Keep in mind, your ideal target client should be the one who would only pay you AFTER you delivered them significant results.

The answer to this question becomes the foundation of your Client Value Map.

Let's use my business as an example. As a reminder, our basic positioning statement is: We help purpose-driven consultants, coaches, and service providers get more (and better) clients without complicated advertising or confusing technology.

The 4 top things we do for clients to get significant results are to help them:

1. Package and simplify their minimum viable magnetic messaging and core offer.

2. Attract more ideal target prospects to sales calls by deploying their messaging to the right client attraction channels.

3. Close and convert more clients using the perfect sales call framework that gets new clients excited to pay up-front and eager to get started right away.

4. Transform new clients into raving fans that actively send qualified referrals by scripting the client experience.

If I can help my clients do these 4 things before, they ever buy from me, this is going to make my company extremely magnetic to my target audience. These are the 4 pillars of my client value map.

Your first step in this process is to answer the question and to define the 3-5 big things you would do to help your clients get the fastest and most significant results for their biggest challenges.

Once you have your top 3-5 value pillars identified, we need to break each of them down into steps to help your clients implement so they can get real results in advance. These steps are what you will teach in your content to build goodwill with your target clients.

Continuing on with my example from above, let's take item #1 from my example: Package and simplify their minimum viable magnetic messaging" and break it down into concrete steps.

To help someone get their minimum viable magnetic messaging defined I would do exactly what I am doing with you in this section of the book. I would have them:

1. Identify their ideal target client with the two simple focus questions.

2. I would have them craft a basic positioning statement for their ideal target client, so they are crystal clear about the main result their ideal client wants to achieve.

3. Then I would teach them how to create a client value map with the frameworks we are covering now.

Once I have these defined, I can create content around these steps and deploy it into the marketplace. This is exactly what I'm doing to promote this book and my company through short videos on Facebook, case studies I'm posting organically in groups on LinkedIn, and podcasts where I'm featured as a guest, all targeted toward my ideal target clients.

You'll want to repeat this process until you have 3-5 smaller steps chunked out for each of your core value pillars identified above. This will become the basis of your minimum viable magnetic messaging.

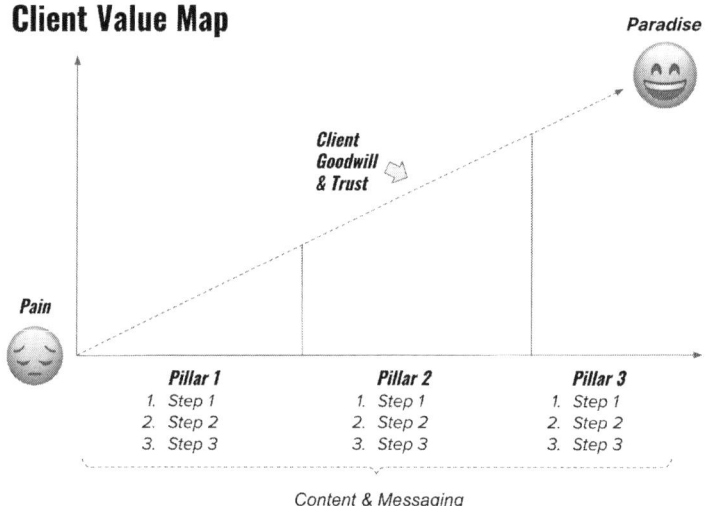

Client Value Map

Example diagram of the client value map with content & messaging pillars.

We've covered a lot of ground in the 'Package Phase' of the Tractionology framework. So far, we've:

1. Gotten crystal clear on your ideal target client with the two focus questions

2. Helped you get laser-focused on your core offer by identifying the stuff you sell that pays you the most money, energizes you, is easiest to deliver, and that you actually enjoy doing

3. Helped you craft your minimum viable magnetic messaging by getting clear on your positioning statement and your client value map

If you've completed each of these exercises, congratulations! You now have the right foundation to set your client-based business up for success in getting more (and better) clients.

Now it's time to move into our next step in your journey where you'll learn how to select the right attraction channels to deploy your minimum viable magnetic messaging into.

SECTION III

THE RIGHT WAY TO ATTRACT AN ENDLESS STREAM OF DREAM CLIENTS

Chapter 8

THE 3 STAGES OF CLIENT ATTRACTION

When I moved to Venice Beach in 2013, one of the first things I did was sign-up for a gym membership at Gold's Gym, aka "The Mecca" of bodybuilding.

From the moment I walked in the front doors, I was confronted with my inadequacy. There were women twice my age lifting 2.5X my body weight. Half the guys in the gym looked like they just got done shooting the cover photo of a bodybuilding magazine.

I had spent the last 14 years as a competitive dancer so I was athletic, but I looked (and felt) like anyone in the gym could pick me up and break me in half like a twig.

Despite the situation and my ego telling me I was in over my head, I mustered up the courage to walk over to the squat racks and begin my workout. After a quick scan of the place, all of the racks were completely full.

It looked like every squat rack was not only full but had people taking turns to share the equipment, with the exception of one rack. There was a guy about my height doing squats with a grey hoodie pulled over his head and giant headphones on. I casually walked up to him, tapped him on the shoulder, and asked if I could work into his set (gym lingo for "can I share this equipment with you").

When he turned around, I had this weird feeling like I knew the guy but couldn't place his face. He looked a little annoyed as he pulled off his headphones and took his hood down and that's when it hit me. There was a very obvious reason why no one else was asking this guy to share the squat rack.

It was because this wasn't "some guy", this was Alex Rodriguez, aka A-Rod, aka the highest-earning major league baseball player of all time. Whoops.

I felt a weird combination of embarrassment, excitement, and moderate humiliation all mixed together at once as I sheepishly repeated my request and added an apology for not recognizing him. He laughed a little, said he just finished his last set, and let me have the squat rack.

When you make the commitment to become world-class at attracting more (and better) clients into your business there is a 100% chance at some point you're going to have your ego tell you that you're in over your head.

There is also a 100% chance you're going to feel your version of "embarrassment, excitement, and moderate humiliation" when you make mistakes and go outside your comfort zone. Push onwards and remember Mental Model #3 "Do the right things and let the results catch up".

With that said, let's dive into Phase 2 of the Tractionology framework, the Attract Phase.

In this phase, you're going to learn how to select the right attraction channel for your business and you're going to learn how to deploy your messaging into the marketplace, get feedback, and iterate until you crack the code of attracting an endless stream of your ideal target clients.

Your main objectives within the Attract Phase, in order of importance, are as follows:

1. **Validated learning.** You want to deploy and validate your minimum viable magnetic messaging with your ideal target prospects to learn exactly what converts.

2. **Book Enrollment Calls.** You, of course, want to get more (and better) prospects booking sales calls or strategy sessions with you.

3. **Build Goodwill.** You want to build goodwill and authority within your market as you test your content and demonstrate that you can help people by actually helping them.

I can't stress this enough, the primary objective here is to learn and validate what works. You're going to make mistakes and that's totally part of the process.

THE ORDER OF IMPORTANCE

For the sake of your sanity and your profit margins, it's important that you don't jump right into paid advertising and automated marketing channels until you've validated your messaging. Validate before you automate. Make it work small, then go big.

When you skip steps and go right into paid traffic, you inevitably spend more money trying to figure out your message. You get a ton of unqualified prospects, you spend hours on strategy sessions and enrollment calls with people who are not your ideal target client, and you accelerate your path to overwhelm, burnout, and frustration.

When you follow the process I'm about to recommend, you're able to craft razor-sharp messaging that attracts and converts your ideal target clients quickly. You're able to make money while you're doing it and if you decide to start buying traffic or doing paid advertising, you're going to be armed with crystal clear magnetic messaging that you know works. Your cost to acquire a client will be far lower, your risk will be lower, and your stress levels will be much lower.

THE 3 STAGES OF CLIENT ATTRACTION

Stage 1 - Get Prescribed: Start with people who (a) you already have goodwill built up with and (b) who have existing authority and goodwill with your ideal target client. The 3 channels I recommend in this stage are client referrals, champions, and partners.

Stage 2 - Go Organic. After you have some momentum and early validation from Stage 1, you're going to go into the wild and start testing your revised messaging with organic outreach. The 3 channels I recommend in Stage 2 are your email list, 1:1 direct outreach, and organic social posts.

Stage 3 - Scale With Paid. By the time you're done with Stage 1 and Stage 2, your messaging will be clear, validated, and magnetic. Now it's time to pour gas on the fire to really scale your outreach. The 3 channels I recommend in Stage 2 are direct mail, partner networks, and paid social / paid search.

One important point before we move on. It's entirely possible to scale to a consistent $100K per month or more with just mastering Stage 1.

Most of the major professional services organizations in the world got to the top without organic social posts or paid advertising (these are billion-dollar organizations). When you truly master Stage 1 and Stage 2, you can earn multiple seven-figures as a "small business" if you stay focused and work through the process.

Before we jump into each of the core attraction channels by stage, I want to help you avoid some of the most common mistakes I see people making when trying to attract more (and better) clients. There are things you're going to need, and there are things you're NOT going to need that will distract you, drain your energy, and waste your time.

THINGS YOU DON'T NEED

1. A pretty website, a fancy funnel, or a shiny logo. Chances are you probably have some of these things already, and if so that's great. If you don't have them, don't worry about them. They are not essential elements on your path to more (and better) clients. If you're pausing to create these assets, you are not focused on the shortest and most direct path to attracting more and better clients.

2. A fancy email autoresponder series or system. Again, if you have this in place already, that's fantastic and will save you time when you get to Stage 3. If you don't, don't

worry about it. Remember we want to validate before we automate.

3. A big budget to buy traffic. I might be beating a dead horse here but it's worth repeating. You don't need any money in Stage 1 or Stage 2. You do need to invest your time and attention in cracking the code and validating your messaging. Make it work small, then go big.

THINGS YOU DO NEED

1. **Courage & Consistency.** You need to show up daily, put the work in to deploy your message, and pay close attention to what is working and what is not working. Take action daily, even if it's something small. Like a flywheel, this process starts slow and builds its own momentum in direct proportion to your moving of the gears and taking consistent action.

2. **Patience, Grace, and Grit**: A lot of what we do with client attraction doesn't work. That's not failure, that's called learning and validation. Be patient with the process, have grace with yourself, and demonstrate grit by not giving up.

3. **A Screening Mechanism**: Once you start to get momentum, you do need a way to screen-out unqualified prospective clients, typically through a survey tool. We're going to cover your sales process and screening mechanisms in the next phase of the Tractionolgoy framework (more on that later).

Attracting your ideal target clients gets easier and easier the more you take action and learn. By far the easiest way to get started is with stage 1 - getting prescribed.

Chapter 9

THE SECRET TO GETTING CLIENTS TO BUY FROM YOU WHILE IGNORING YOUR COMPETITION

In 2020 I got an ear infection and had to go to the doctor during the middle of the pandemic and widespread lockdowns here in Los Angeles. I had been spending more and more time on zoom calls and was using my Apple Airpods as my go-to microphone and audio choice. Turns out, wearing Airpods for hours a day and not cleaning them is a great way to get a bacterial ear infection.

After a quick visit with the doctor, he gave me a prescription for these special ear drops called Ofloxacin and placed an order for a pharmacy near my house. This is where things got interesting.

I walked into the pharmacy, walked all the way to the back, and got in line so I could give my prescription to the pharmacist. There were 3 people ahead of me and after a 15-minute wait, I gave the pharmacist my name and she told me I had to wait (again) while they prepared the prescription.

After another 10-minute wait, they called my name and I finally got my prescription. I walked back through the aisles in the store and went home.

You might be wondering, what the heck is interesting about this Joe? This sounds like every other trip anyone has ever made to the pharmacist to get a prescription. What does this have to do with getting more (and better) clients?

Let's rewind the situation and breakdown the details to explain why this has everything to do with getting more (and better) clients for your service-based business.

First, it started with my trip to see the doctor.

- **Authority & Credibility**: My doctor's office is full of signals that he is the authority. The degrees on the walls in the office, the white uniform, the people working for him. These all create deep conditioning with his patients that have programmed us to think he is the expert, and we should trust him.

- **Goodwill**: I have been to this doctor before, we have a relationship. I trust him. He knows me. Specifically, he knows I'm skeptical of traditional medicine and that I very rarely call in for help with anything unless it involves an emergency or giving birth. So, he only prescribes me medicine when he absolutely has to.

- **Specificity**: My doctor diagnosed my situation with a lot of qualifying questions before he prescribed me anything and when he gave me the prescription it was a solution that was designed to solve my exact pain.

Then, it continued with my trip to the pharmacy:

- **Option Ignorance**: As I walked to the back of the pharmacy, through the endless aisles of other drugs, I completely ignored and disregarded the dozens of solutions on the shelves that could have solved my problem. It was as if they didn't exist.

- **Experience Tolerance**: Having to walk to the back of the store, wait in line, then wait again to buy something is a terrible experience. I can literally order anything else on the internet and have it sent to my house in 24-72 hours with Amazon, but I canceled meetings and rearranged my day to get my prescription and didn't even think twice about it.

- **Price Inelasticity**: I didn't even look at the price of the drug and if it had another zero attached to it, I would have paid for it anyway . I was in pain and fully believed this was the thing that was going to get me out of pain.

Wouldn't it be great if you could get "prescribed" by people who had authority and trust with your ideal target clients in a way that got them to buy from you while ignoring all the other available options and then not haggle with you on price? That is exactly what Stage 1 of client attraction is designed to teach you to do.

This isn't a theory I read about and regurgitated for you to sound smart. I've been using this for over a decade to get high-profile, high-paying clients to buy from me again and again.

In 2010 I was doing independent consulting and was looking for another big client to take on. I emailed a friend of mine using the exact techniques I'm going to show you. This friend recommended me as the consultant to do a project for the head of Microsoft's Server & Tools division. The main sponsor for the project was a super sharp guy named Satya Nadella. In fact, he's so smart they later made him the CEO of the company.

In 2014 I had just bombed my first technology start-up and launched a digital marketing agency. The agency wasn't producing the cash I needed to survive so I emailed a handful of trusted mentors and explained my situation. I asked if they would serve as 'champions' on my behalf to introduce me to potential clients that I might be a good fit for. One of my mentors also happened to be a trusted advisor for a guy named Rob Dyrdek. A few weeks later, I was working for Rob to help him craft the strategy and operations to support a new investment fund he was launching. Rob is a celebrity entrepreneur with a seemingly infinite number of people he could have hired to help him with this, but I quickly got the gig after being prescribed to him by my mentor.

In 2019 after my son was born, I decided to take a break from the tech industry and launched the consulting and coaching practice that would eventually become The Tractionology Group. I'll bet you'll never guess how I got my first 10 clients? I got other people to prescribe them to me.

In all 3 examples, there were no proposals to send or bidding wars with other service providers and none of them told me they needed time to think about it or that I was too expensive.

Just like my doctor prescribed me my medication, I was able to get other people to prescribe my services to my ideal target clients.

Now it's time for you to do the same for your client-based business using 3 simple attraction channels.

CHANNEL 1: CURRENT AND PAST CLIENT REFERRALS

This is by far the easiest channel to test your messaging and get some quick momentum to grow your client-based business.

I'm shocked by the number of client-based businesses that don't have at least two processes in place to systematically ask for referrals.

When you look at the data, referrals typically close faster, have higher lifetime customer value, and are more likely to refer you again if you can deliver results. Referrals are very powerful.

Because these people have already done business with you and received results from your services, you likely have lots of goodwill already built up with them. All you need to do is make an ask and make it easy for them to refer you to your ideal target client.

Here are some rules of the road when asking for referrals:

1. Make a clear ask. Make it clear that you are asking them for help and if you offer a referral bonus of some kind make that clear too.

2. Tell them exactly what to do and make it easy for them. I like to make the emails I send asking for referrals "for-wardable" or to include an email they can easily copy and paste when making a warm introduction for me.

3. Use your positioning statement and be clear about the value you provide. Make it easy for everyone to understand exactly who you help and the results you deliver.

I have included 2 different referral request email templates in the digital supplement provided with this book. These are actual emails that have been used repeatedly and gotten me results. You can leverage these as a starting point for your own referral outreach.

You can get access to these template referral requests by going to www.tractionologybook.com/resources.

CHANNEL 2: CHAMPION REFERRALS.

Champions are people who know you, trust you, love you, and want to support you and the work you do.

You don't have to pay these people to help, they just want to help you because of the existing relationship you have with them.

What's great about champions is you can leverage the good-will they have with your ideal target clients without much effort from everyone involved and it actually benefits all parties involved if you're doing good work and actually helping people with what you offer.

I know eight-figure coaching companies that built their entire book of business from relying on champion referrals as their only client attraction channel. This channel is easy to get started with and by far the best way to get 'prescribed' to dozens of your ideal target clients.

To get started with champions here's exactly what to do:

1. Make a list of every possible person who could be a champion for you. A good champion (a) has a great relationship

with you and (b) has a great relationship with your ideal target clients.

2. Reach out to these people one by one and get them on the phone or meet them in person if possible. You don't want to just email them and ask for referrals. You want to speak with them so you can enroll them in your vision, and get them bought into who you are serving, the problems you will solve for your ideal target client, and the impact you want to have.

3. After meeting with your potential champion, send them a thank you email that they can easily forward to people they know who fit your ideal target client description.

4. Promptly respond to any introductions your champions make and be proactive in getting things scheduled with any contacts they send your way.

5. Do good work and demonstrate you can help the new contacts by actually helping them before you ask for the sale.

CHANNEL 3: REFERRAL PARTNERS

Think about all the people your ideal target client buys from before, during, and after they buy from you. These people are spending their hard-earned time and money to build goodwill with your ideal target clients and many of them have incredible influence and authority over them. You want to partner with these other service providers to leverage their influence and get you 'prescribed' to your ideal target client.

Let me give you a real-world example. I work with a successful consulting and coaching company based in Western-Europe. They help ambitious six-figure small businesses scale to multiple seven figures through an immersive year-long group coaching program.

I took them through an exercise to help them identify potential referral partners who also sell to and have authority with their ideal target client, and we identified dozens of possible options. One of them was an advertising agency that serves the same demographic and has tremendous authority and goodwill in the marketplace because of how well they deliver client results.

My client reached out to this agency and they agreed to partner in exchange for a reasonable revenue share. My client offered the agency's existing clients a free 45-minute, pure value, training using the content they created from their Client Value Map and was able to close €100K in new business in 72 hours from this one presentation.

To get started with referral partners, here's what to do:

1. Make a list of every possible service provider or business that sells to your ideal target clients before, during, or after they buy from you. Don't discriminate for any reason, even if the people on the list are your direct competition. Just build the list.

2. Rank the list based on your existing relationships with the providers on the list so that the ones you have the strongest relationships with are at the top, and the ones you have little to no relationship with are at the bottom.

3. Craft a short email to each referral starting at the top, working your way to the bottom. The objective of the

email is to get these folks on the phone so you can enroll them in your vision and get them excited about how you can add value to their clients or customers while making them some extra money at the same time.

4. When these partners start saying yes, work with them to schedule a free training for their audience on Zoom or Google Hangouts.

5. Use your Client Value Map to demonstrate you can help the audience by actually helping them and close with a clear call to action to book a strategy session with you. Don't get caught up trying to execute "the perfect webinar" or "video sales letter" or any of that. Just train, add lots of value, and make it crystal clear you want them to sign up for a strategy session at the end. You can iterate the delivery of this training over time and test what messaging connects the best with your ideal target client.

For many of my clients, a well-executed Stage 1 is all that is required to scale their client-based business to consistently generate $100K or more per month in revenue. If you can get people to prescribe you consistently, you're in a powerful position of growth.

If you want to grow even faster and add a second pillar of stability to your business, you're going to want to move on to Stage 2, where you'll learn how to professionally generate hundreds of qualified leads a month using the 3 core organic outreach channels.

Chapter 10

HOW TO DO DIRECT OUTREACH WITHOUT BEING SPAMMY

I like to think of income sources for my business like legs under a table, where the more of them I have, the more stable everything becomes.

Such is the case with Stage 2 where we're going to start communicating with and influencing our ideal target clients directly in the marketplace to build a second leg or income source to further stabilize your client-based business.

Personally, this is my favorite stage of client attraction because I love selling. I love being of service, creating value for people one-on-one to generate goodwill from thin air, and then translating that goodwill into enrolled clients who get transformative results.

There is no better place to deploy, validate, and iterate on your minimum viable magnetic messaging than through direct-to-client organic conversations.

CHANNEL 4 - EMAIL

Surprisingly, one of the next most underutilized assets I see my clients neglecting is email. They have done business with, interacted with, and captured contact information from thousands (in some cases hundreds of thousands) of people from all over the world and they leave this "asset" completely underutilized.

If you have an email list, or if you can quickly create one from all your past contacts, this is one of the lowest hanging fruit opportunities that you have right now and it's also one of the fastest ways to validate your messaging and content while earning some quick cash.

One of my clients runs a consulting business that helps companies in the oil industry improve their employee engagement through easy-to-implement training and leadership development seminars.

They've been in business for more than 20 years and have collected thousands of emails from prospective clients who they met at tradeshows, networking events, mixers, and so on. I took them through the same exercises we covered in the "Package Phase" of this book to help them craft their minimum viable magnetic messaging.

Once we had their messaging and content drafted, I asked them if they had an email database and they said what most clients say "yes but we don't do anything with it", which was music to my ears.

I helped them quickly categorize and segment their list and prepare a really short series of emails to send out with an offer to attend a free 45-minute training session. We sent the emails out and they got 70 people to attend their training where they taught content identified from their Client Value Map. Of those 70 attendees, 8 people booked sales calls and 4 people became new clients. In less than 2 weeks they took an asset that was sitting right under their noses and converted it into $80,000 in cash collected with another $160,000 in accounts receivable.

Here's how you can replicate this process for yourself:

1. Segment your email list to identify the prospects that most closely match the ideal target client you identified in the Ideal Client Focus exercise from early in the book.

2. Sign up for a video broadcasting tool that allows you to easily register attendees. My go-to tool is Zoom but if you don't want to pay for the premium version you can use Google Hangouts for the video conference and a simple Google form for registration (both of these are free and incredibly easy to use).

3. Write a brief but compelling email to your list, inviting them to attend a free training where you are going to demonstrate you can help them by actually helping them.

4. Send the email 5 days in advance of the free training and keep sending the email to everyone who doesn't register for the training once per day for the next 3 days. Make sure you change the subject line slightly for each new email to test different angles that get them interested enough to open the email.

5. Host the free training and invite attendees to join a free strategy session or sales call at the end.

6. For everyone who did not attend or who attended but didn't sign-up for a sales call, send them a follow-up email with a recording of the session.

7. If you have the phone numbers of the attendees who did not attend, call them and let them know you created a training specifically for them that you'd like them to see, then either host another session and practice your content again or send them the recording.

CHANNEL 5 - DIRECT OUTREACH

By far, Direct Outreach is my personal favorite client attraction channel. It's old school, it's 1:1 sales, and it's incredibly effective at giving you rapid feedback on your messaging.

If your minimum viable magnetic messaging is truly magnetic, people will engage with you. If not, they won't or they will in many cases simply tell you to go away. It's one of the best entrepreneurial laboratories on the planet to deploy, track, and iterate on your messaging.

By implementing a daily regimen of direct outreach to your ideal target clients, you can generate consistent leads for your business while lowering your customer acquisition costs because it doesn't cost you a dime to reach out and connect with people on a 1:1 basis.

Three of my favorite platforms to deploy a direct outreach strategy into are Facebook, LinkedIn, and Instagram because on these platforms it's easy to connect with people and it's incredibly easy to start a conversation with them.

Here's how to get started with direct outreach:

1. If you're connecting with people on a social platform like Facebook, LinkedIn, or Instagram, make sure your profile is cleaned up and has your positioning statement clearly articulated upfront. This lets people know who you are and how you can help.

2. For Facebook and LinkedIn, join groups that your ideal target clients are likely to frequent and then request to join these groups.

3. Once you're admitted into these groups, look for recent posts from group members and start paying attention to who might be your ideal target client. You can also click the members list on Facebook or LinkedIn to see the entire list of members in that group and start researching who might be a good fit.

4. When you see someone that you think is your ideal target client, send them a friend request, connection request, or follow them (depending on which platform you're using). Aim for 30-40 connection requests a day and be mindful of any limits the platform you're using might put on the number of new connection requests they allow you to send.

5. Once they've accepted your connection request, send them a well-crafted message to see if you can help them.

6. If the person responds, you want to offer to get on a call with them to build rapport and qualify them further. If you're confident they're a good fit, offer to do a strategy session where you can help them get clarity and solve a

real problem in their business. We will cover exactly how to conduct a strategy session or enrollment call in great detail in the "Convert" section of this book.

Here are some best practices with direct outreach:

- Remember that you're talking 1:1 to another human, personalize each message, and don't use a script. People can smell that a hundred miles away and it strips you of your authenticity.

- If you're going to DM (direct message) people on Facebook, LinkedIn, or Instagram, make sure you send a connection or friend request first, so you don't end up in their spam folder.

- When you join groups on Facebook or LinkedIn don't just jump in and start posting your offer and spamming everyone. This will usually get you kicked out of the group.

- Block time on your calendar to do direct outreach daily. I like to break my time into (a) finding new groups, (b) observing these groups and finding people who might be a good fit, and (c) connecting with and sending messages to 30-40 new people a day.

- Test different versions of your message to see what resonates and what does not. This is an excellent channel to quickly test your messaging.

For an example of a simple message, you can send to start conversations with your ideal target clients through direct outreach, you can see several real-life samples I still use today by going to tractionologybook.com/resources.

CHANNEL 6 - ORGANIC SOCIAL POSTS

Organic social posts are simply another form of direct outreach. The rules are mostly the same with some nuances depending on the platform you're using.

Social platforms have a funny way of changing their user experience the moment you try to document how they work in order to tell someone how to use them. So rather than go through one specific platform, I'm going to give you a high-level overview of how to use organic social posts as a client attraction channel that you can adapt to any platform.

Here's how to get started:

1. Make organic posts letting your followers know about your success and the success of your clients. If you have case studies, share them.

2. On your post include a call to action for anyone who wants more information to leave a comment on the post.

3. Send a friend or connection request to each person that comments or likes the post and begin the direct outreach protocol above to get them on a strategy session. Make sure you only friend request people who appear to be your ideal target client.

4. If you don't have any success stories to share or you have run out of them, post other relevant stories, questions, insights, tips, and content related to your client value map.

5. Include a call to action on each post and follow up with everyone who engages that fits your ideal target client profile.

With organic social posts, here are a few guidelines to keep in mind:

- **Clean up your profile**. Add your positioning statement to your profile so target clients know how you can help them.

- **Be consistent**. Post content often, daily if possible. Even if no one is responding or commenting when your target client finds you, the first thing they will do is binge review all your posts.

- **Create value**. Make sure when your ideal target clients get to your page they aren't looking at what you ate for dinner and instead are being delivered value from case studies and content from your client value map.

- **Respond & engage**. Interact with the people who comment on your content. The idea is to be 'social' so make sure to engage and follow-up.

- **Be human**. Don't just post business content. Marketing is human to human, heart to heart. Your ideal target clients want to know you're a human being outside of your business. So include personal posts that demonstrate your values, especially the values you hold that your ideal target client also holds.

- **Be patient**. It takes time to build up traction with this channel but it can be an incredibly effective and free mechanism to get you more (and better) clients.

Chapter 11

WHY STARTING WITH FACEBOOK OR YOUTUBE ADS IS A RECIPE FOR DISASTER AND WHAT TO DO INSTEAD

Before we go any further there are a few things you need to have in place to cost-effectively scale your messaging with paid advertising:

1. Your minimum viable magnetic messaging and content from your value map should be validated and confirmed by deploying it into the marketplace, talking to your ideal target clients, and generating qualified leads.

2. You should have collected credit cards or cash or payment of some kind as evidence that your offer is valuable and resonates with your ideal target clients.

If you don't have a validated message and a proven offer, you're going to have a very difficult time turning paid advertising into profit.

Simply put, you're going to spend a lot of money and a lot of time sending attention and eyeballs to messaging and an offer that does not convert.

If you're trying to create a recipe for failure, doing paid advertising without a validated message and a proven offer are the two essential ingredients.

If on the other hand, you do have razor-sharp messaging that turns prospects into qualified leads and you have a proven offer that has converted into cash, then you're ready to begin your journey of using paid advertising to get more (and better) clients.

When properly executed, scaling with paid advertising is truly like adding gas to a fire in terms of the effect it can have on your marketing and sales. As with getting prescribed and with implementing organic outreach, I recommend you make it work small and then go big.

The 3 channels I recommend you start with for paid advertising include direct mail, partner networks, and paid social (e.g., Facebook, Google, LinkedIn).

CHANNEL 7 - DIRECT MAIL

Direct mail is one of the most underutilized attraction channels on the planet. It simply involves sending a physical piece of mail through the postal service to someone you have identified as your ideal target client with an offer to connect with you.

In the age of email, direct messages, text messages, WhatsApp messages, and in-app messaging, most entrepreneurs have forgotten about direct mail, making it an easy channel to get attention with because it's significantly less crowded.

What's great about direct mail is that it allows you to cost-effectively zero in on and target a specific person, it feels personalized, and when done correctly captures your ideal target client's attention long enough to see your offer. It demonstrates to your ideal target client that you're serious and helps you stand out from all the other service providers in the marketplace.

There's one small twist to this channel that you need to understand. You're not going to send regular letters, you're going to send "lumpy mail" or packages that include something inside of them. This helps the package standout, increases deliverability, and makes it more likely the intended recipient will actually open the package.

One of my favorite marketers and business strategists, the late Chet Holmes, used a combination of lumpy mail and direct outreach phone call follow-ups to double sales three years in a row for a magazine owned by billionaire investor Charlie Munger (Warren Buffet's business partner).

In 2019 I had a client who ran a software and services business that was selling into medium and large organizations with hard-to-reach decision-makers. We followed the process I'm about to show you to help him generate over $250,000 in sales in 8 short weeks.

Here's how to get started with direct (lumpy) mail:

1. Make a list of your ideal target clients, preferably in Excel or Google sheets.

2. Research each client on google to find their mailing address or shipping information.

3. Write a short letter that tells the prospect who you are, what problems you solve or results you get, a personalized note about how you think you can get results for them (use your positioning statement), and a call to action for them to get on a strategy session or attend a personalized training session (where you help them using your client value map content).

4. Include a small gift or trinket in the package, this is what makes it "lumpy". For example, we sent "sushi socks" (socks that are folded to look like sushi) to prospective clients and the call to action in the letter was to get real sushi or lunch of their choice on us.

5. Send the lumpy mail packages using your preferred mailing service.

6. Repeat this process with a new letter and new small gift twice a month until you get a response.

Download example direct mail letters and see example gifts you can use in your own direct mail campaigns by going to tractionologybook.com/resources.

CHANNEL 8 - PARTNER NETWORKS.

Remember the list of other businesses that your ideal target client buys from before, during, and after they buy from you that we discussed in Channel 3 - Referral Partners? We're going to revisit that list and leverage a slightly different strategy to engage these partners.

Rather than incentivizing your potential referral partners with a revenue share for any prospects that enroll as clients, we're going to pay these partners upfront to promote you to their list of clients instead. The key difference is now we have validated your messaging to ensure that it's absolutely magnetic and converting so there is less risk for you to make this investment.

I recently helped a client create and deploy a partner network strategy that doubled his revenue in less than 8 months. This client, "Dale", helps corporate managers successfully transition into becoming six-figure entrepreneurs in six months through a group coaching program.

Dale's business was generating about $60,000 a month in revenue and he wanted to scale up to consistently generate $100,000 or more a month. After working with Dale to package and validate his messaging, we made a list of the top 50 potential partners who were publishing on the networks where Dale's ideal target clients frequented. These included places like online forums, private member organizations, and select social media platforms.

Of the 50 target partners, 22 responded to Dale's outreach, and 8 ultimately agreed to promote Dale and his business to their audience. From those 8, Dale was able to create custom content that added value, generated goodwill, and converted his ideal target prospects into happy clients.

Here's how you can get started with partner networks:

1. Make a list of every possible service provider or business that sells to your ideal target clients before, during, or after they buy from you. Don't discriminate for any reason, even if the people on the list are your direct competition.

Just build the list or update the list you created from reaching out to referral partners in attraction channel 3.

2. Craft a short, personalized email to each potential referral partner that is customized, demonstrates that you researched them and their business, explain how you can create results for their audience, and include some of your top accomplishments or case studies so they know you are real. Include a clear call to action. The goal is to get them on the phone or meet in person to discuss and finalize the partnership.

3. When you get connected and agree to deal terms, you can follow a similar strategy outlined for referral partners where you hold a group training that creates goodwill with the content from your Client Value Map and ask the group to book a strategy session with you.

CHANNEL 9 - PAID AD NETWORKS.

Giving you a detailed guide on how to run paid traffic on each of the major ad platforms is beyond the scope of this book.

With that said, most of our clients who run paid ads are successfully getting a positive return on ad spend (ROAS) on the major ad platforms because they follow a set of rules to keep them from losing their shirt on advertising costs.

These rules are as follows:

1. **Define your goals, timeline, and budget**. Before you spend a dime in paid traffic on the major ad platforms, spend some time to define the targets you want to hit, how

long you're willing to play the game to get those targets, and exactly what you're willing to spend to pull this off. This will serve as a north star and 'throttle' on your daily ad spend.

2. **Define your sales process**: Get clarity on the journey you will take prospects through as you generate goodwill and convert them into paying clients. I strongly recommend keeping it as simple as possible to start until you have validated that your ads are converting into paying customers before you get fancy with automation, email autoresponders, and complicated funnels.

3. **Test your sales process**: Make sure to run a test prospect through the entire sales process to double-check everything is working correctly, emails are being recorded where they are supposed to, pixels are firing correctly, your application service is correctly getting responses, and your scheduling service is getting meetings on your calendar. I can't tell you how many times I've sat a client down who is running paid traffic, made them re-test their process, and helped them find hundreds of leads that got "stuck" somewhere in the process because they didn't test properly. The simpler your sales process, the less complexity you will have with testing.

4. **Track your progress daily**. All of the major ad platforms have their own dashboards and report-outs to help you understand if you're getting traction and sadly most of them are incomplete or don't track everything you need. I recommend setting up a very basic spreadsheet that tracks the data of how your leads are progressing through your sales process. Some of this data you will get from the ad platforms and some will come from other sources

depending on your sales process. It's imperative you track your numbers to understand if what you are doing is working. This will help you cut your losers and to let your winners run.

5. **Mind the "cash gaps"**. The space between each step in your sales process represents a potential cash gap. If you can advance people through to the next step, you will increase your chances of getting them in front of your offer and converting them into a paying client. For example, if someone clicks your ad and is taken to a registration page to watch a free training video and they don't register, that's a gap. If they register but don't watch your video, that's a gap. If they watch but don't watch long enough to see your offer that's a gap. If they see your offer but don't book a call with you, that's a gap. Make sure you have an automated or manual mechanism in place to follow-up with people who get stuck in the gaps. Most people who watch your ads will not convert all the way to book a strategy session. Give them a little nudge with strategically placed email follow-up and ad-retargeting.

6. **Make it work small, then go big.** Even though you will have validated your messaging through organic channels, that's no guarantee of success when you run paid traffic, it simply de-risks your marketing and gives you a head start. This is why I recommend all my clients start small, gather data, iterate and improve each step until all of their conversions through the sales process are within the target performance indicators, they set in rule #1 above. Once they are performing at the desired conversion levels, then it's time to scale ad spend.

7. **Play the long game**. If you have the budget and stay committed, you will win. Far too many business owners give up within the first 30-45 days because they aren't seeing the results they want. The objective here, as with organic traction channels, is first and foremost validated learning then conversions and cash. The reward for getting this channel optimized is more (and better) clients than you will be able to handle, so hang in there and keep validating, learning, and iterating until you crack the code.

If you've successfully validated your minimum viable magnetic messaging and want help specifically with running paid advertising on Facebook, Instagram, Google, YouTube, or LinkedIn, reach out to my office so we can get you some additional resources (http://www.tractionologygroup.com/apply).

Chapter 12

THE NEW SCIENCE OF FINDING YOUR MESSAGE MARKET FIT

Now that you've gone through the 3 stages of client attraction and the 9 attraction channels, it's important to remember that the number 1 objective in this phase of the framework is to get validated learning from the market so you can optimize your minimum viable magnetic messaging.

In other words, we need to make sure you have what I call "message market fit". You want to transform your messaging from an educated guess to a battle-hardened, market-validated, razor-sharp masterpiece that generates goodwill in your market and gets your ideal target clients excited to work with you.

You don't want to leave this up to chance or guesswork. You're not going to throw every message out there and see what sticks. There will be no "spray and pray" messaging going into the market.

Instead, you want to use a specific process to scientifically deploy, test, and iterate your messaging until you've validated that it's compelling to your ideal target client.

To pull this off, you need to follow the 6 steps of the Message Validation Process.

MESSAGE VALIDATION PROCESS

1. **Select**: Pick the attraction channels you want to start with based on your evolution as a client-based business. If you haven't perfect stage 1 (getting prescribed) I recommend you start there.

2. **List**: Make a list of all the past clients, all the potential champions, referrals, people on your email list, etc. that you plan on reaching out to.

3. **Deploy**: Send version 1 of your minimum viable magnetic messaging through the attraction channels you've selected. I recommend segmenting your list into batches of 30 so you can deploy messaging and track results, then adjust if needed before you send the next 30, and so on.

4. **Track**: Keep track of your progress by channel. If you send 10 referral requests, how many responded? How many sent referrals? If you segmented your email list and sent an email, how many opened? How many clicked through to the next step? Use a spreadsheet to track progress, it doesn't need to be fancy, but you need to keep track.

5. **Iterate**: If you didn't get the results you wanted, iterate the messaging or the medium on the next deployment. It's important you only change one thing at a time so you can isolate what changes actually drove the results. For example, if no one is opening your emails, don't rewrite the message or change the medium (e.g., from text to video),

change the headline only. If the email got opened, consider tweaking the messaging or changing the medium from text only to video or add an image, etc.

6. **Repeat**: Deploy your updated messaging, track the results, iterate as needed. You don't need fancy software for this, you just need the discipline and determination to keep testing and iterating until you crack the code.

This brings us to the end of the "Attract Phase". We covered a ton of material and the goal with this section was to arm you with some powerful and actionable guidance you can use to validate your messaging, get more clients, and build goodwill in your market.

Now it's time to turn our focus to the third step in the Tractionology framework where we'll focus on transforming warm leads into paying clients with a powerful sales process and the perfect sales call framework that will get your target clients excited to pay you up-front and start working with you right away.

SECTION IV

HOW TO POWERFULLY ENROLL YOUR DREAM CLIENTS

Chapter 13

HOW COMPLEXITY SLOWLY KILLS SERVICE-BASED BUSINESSES

In the summer of 2020, as the COVID-19 pandemic-driven lockdowns began to temporarily lift, "David" was eager to get his coaching practice back into growth-mode. Prior to things shutting down in the U.S., his coaching practice was generating about $70,0000 a month but growth had flat lined during the peak of the pandemic.

When David and I started working together in August, a familiar pattern began to emerge. David had multiple offers, multiple client avatars, multiple sales processes, a complicated web of email automation, and he was completely reliant on paid Facebook traffic as his only lead source.

David and his team were completely overwhelmed, they couldn't figure out why their ad costs were going up and their close rate on sales calls was going down.

David confided in me that he was afraid to let any of his team go during the pandemic and that as his ad costs went up, he began to worry about not having enough cash reserves to weather the storm if things got much worse. If he didn't figure something out fast, he was going to have to start firing people on his team.

While taking David through the Tractionology framework, the first thing we did was to get his team focused on their core offer and their ideal target client and to ignore everything else. This immediately gave David and his team some breathing room, took away their overwhelm, and allowed them to simplify and focus their energy on their best offer and best clients.

As we dove into the rest of the framework 4 common problems began to emerge with David's core offer:

1. **Wrong Messaging**: The messaging in his ads, landing pages, and email follow-up was completely out of touch with the sentiment of the market.

2. **Complex Process**: His sales process was incredibly complex and difficult to iterate because of all the automation he built into it and the process was no longer creating value for his ideal target client at each step and interaction.

3. **No Filtering**: As David and his team got more desperate to grow the business, they stopped filtering out leads who were not a good fit for his core offer and let nearly anyone with money get on a sales call.

4. **Undisciplined Calls**: David's two-person sales team was no longer following any kind of process to convert prospects into clients, David (their best closer) was no longer taking sales calls himself, and no one was monitoring the patterns on the sales calls to understand what clients were actually looking for.

Fortunately for David, I've seen these patterns over and over again and I had a pretty good idea of what he needed to do. I took him through the 4-step framework I'm about to show you and with a bit of elbow grease and a ton of courage, David and his team were able to break past $110,000 in revenue in just under 60 days.

OBJECTIVES OF THE CONVERT PHASE:

- **Continue Learning**: Your sales process and sales calls are by far the best place to continue to validate and polish your messaging. Remember the primary goal is to crack the code with continuous validated learning.

- **Generate Cash**: At this point in your journey, you'll be able to apply the frameworks in this section and start converting warm leads into paying clients.

- **Simplify & Focus**: Chances are you have a sales process or more than one sales process you've used to enroll clients into your business. We're going to help you find your best process, simplify it, and then optimize it so it generates goodwill for your ideal target clients and helps you close more sales.

With that said, let's jump into the framework I used to help David and his team get more (and better) clients.

THE 4 STAGES OF THE CONVERT PHASE:

1. **Evaluate**: Evaluate your current sales process to make sure it is optimized for (a) creating goodwill for your ideal

target clients through each step and (b) is hyper-focused on getting the right clients on sales calls that don't require endless follow-up.

2. **Exclude**: Strategically filter out prospects who are not your ideal target client with closely monitored screening criteria.

3. **Engage:** Leverage carefully selected touchpoints to engage your ideal target clients to ensure they show up to your sales call ready to engage.

4. **Enroll**: Powerfully enroll your ideal target clients with the 'perfect sales call' script.

Optimizing the conversion of your sales process is one of the most effective ways to immediately get more (and better) clients into your business.

The more effective your sales process, the fewer leads you need to attract to grow your client-based business and the more time you can spend serving clients, making an impact, and transforming clients into raving fans that send you a mountain of referrals.

Chapter 14

THE IGNORANCE TO OVERWHELM TRAP AND HOW TO AVOID IT

Over the last decade, I've seen an interesting trend emerge in the coaching, consulting, and service-based business industry when it comes to enrolling and converting new clients. That trend has been the transition from ignorance to overwhelm.

10 years ago, the majority of coaching, consulting, and other service-based businesses didn't know what a sales funnel was, they didn't have an email autoresponder, they didn't know how to run paid traffic or how to run automated ad retargeting. All of these tools existed, but the service-based business industry was largely ignorant of them.

Fast forward to 2021 and the exact opposite is true. Just about every business owner in the space has a website, a sales funnel, a CRM, an autoresponder, automated scheduling services, software to write their copy, and on and on. What I see now is a lot of service-based business owners who are overwhelmed with 'too much stuff'.

Don't get me wrong, I love technology and automation. You can pop up a 5–7-person company, a high-ticket offer, and scale to $10M+ in revenue with no outside capital in a few short years and many service-based businesses have done exactly that.

The trouble is the vast majority of business owners get stuck "majoring in minor things" as Jim Rohn would say. They get caught in the tech and automation and lose their read on the market and lose touch with the impact they originally set out to have on their clients in the first place.

To avoid this trap, you simply need to evaluate and simplify your sales process so you can stay focused and make small changes to optimize each customer touchpoint.

STAGE 1: EVALUATE

In the 'Evaluate Stage', the goal is to help you simplify your sales process so you can focus on creating goodwill for your ideal target clients at every touchpoint while continuing to validate your messaging.

Depending on your industry, your sales process may look a bit different but for most service-based businesses here is the sales process I recommend you leverage to simplify and focus:

1. **Lead**: Someone sees your minimum viable magnetic messaging and content from your client value map and raises their hand to talk to you or you hunt them down using direct outreach.

2. **Apply:** They fill out a short application or survey before you talk to them. This gives you powerful data to

customize the sales call or strategy session to fit their specific needs.

3. **Call:** They get on a sales call or strategy session with you where your goal is to actually help them, not to sell them or close them or manipulate them into buying.

4. **Close:** If the person is someone you can actually help, they want to get help, and they specifically want your help, you powerfully enroll them to become a client.

That's it. No endless follow-ups or retargeting (yet) and no shenanigans. This simple and honest approach works for coaches, consultants, agencies, online course creators, insurance agents, wealth managers, attorneys, creative freelancers, daycare centers, private schools, and so on.

By focusing on these 4 simple steps, you can get out of being overwhelmed, and get back to cracking the code on exactly what is needed to optimize your sales process for more (and better) clients.

Chapter 15

FILTERING OUT THE NOISE

Before you get on the phone with a potential client, we need to make absolutely sure that they:

1. Are in fact your ideal target client.

2. Can afford your services.

3. Are interested in working with you.

By filtering out the people who are not a good fit (e.g. people who don't fit these 3 criteria), you're going to save a lot of time and avoid a ton of frustration.

Remember our Ideal Client Focus Question from the Package Phase?

If you could only get paid AFTER your client got the results you promised, what relevant characteristics would they have to have in order for you to take them on?

We only want to talk to these people, we want to filter out everyone else. This respects your time and the time of those who are not a good fit.

We also only want to get on the phone with potential clients who can afford to work with us right now. If your goal is to enroll

8 clients this week, would you rather do that by taking 40 sales calls or by taking 10?

Lastly, if your messaging is doing its job by creating goodwill with your ideal target clients, they should already be interested in working with you before they talk to you. The call or strategy session will simply add even more goodwill making it a no-brainer to do business with you.

STAGE 2: EXCLUDE

To filter the wrong people out and to collect valuable information from the right people, I recommend you use a simple survey tool (e.g., Jotform, Google Forms, Typeform, etc.) to collect the following information:

- **Contact Info**: First name, last name, email, phone

- **Website**: Applicable if you sell to other businesses.

- **Baseline Question**: How much revenue are you currently generating each month? You want to ask a question that gets the ideal target client to baseline their current situation as it relates to the results your positioning and messaging have been helping them to improve. If you're coaching people on weight loss, you might ask "How much do you currently weigh" or if you do mindset training to help people reduce their anxiety you might ask "On a scale of 1-10, how would you rate your current daily anxiety level", etc.

- **Desire Question**: How much monthly revenue would you like to be generating? Now that you have a baseline, you want to know what their ideal outcome is. Similar to the baseline question you'll want to customize this based on the results you help your ideal clients achieve.

- **Obstacle Question**: What do you feel is your biggest obstacle to hitting your revenue goal? You want to understand what the prospective client thinks is blocking them from achieving their desired situation. You can replace 'revenue' with anxiety, weight loss, or whatever the result your baseline and desire questions focused on.

- **Situation Question**: Briefly describe your business, who you serve, what you sell, and the price point you offer. Here you want to understand their current situation to get some additional context. If you were doing weight loss you might replace this with "Briefly describe your diet, exercise habits, and overall personal health" or something similar.

- **Interest Qualification**: We offer 1:1 consulting to help service-based businesses get more (and better) clients without complicated ads or confusing technology. Is this something you're interested in learning about on our call? This question gets right to the point and asks the person in advance of the call if they have an interest in working with you. If the answer is 'no' they aren't ready to talk to you and you should filter them out. You will want to customize this question using your positioning statement.

- **Financial Qualification**: How willing and able are you to invest in growing your business right now? This question is a proxy indicator for their willingness to invest in finding a solution. If someone answers this question with "nothing" or "none" then you should not meet with them. Again, customize this to fit the context of your business.

- **Urgency Qualification**: If you're accepted as a client, how soon can you get started? This is a qualification question that helps you understand how willing and ready this person is to work with you.

- **Confidence Qualification**: Why is now the time to grow your business? This is a powerful question to understand if the person applying is ready for change. It causes them to pause and reflect on their desire to make a shift in their lives and to do the work. Again, customize this to fit the context of your offer.

After reviewing the results from the application if you find out that a person is not a good fit, simply send them a quick note explaining that you will not be moving forward with the meeting. Explain that based on their answers you've determined they aren't a good candidate for your services, and you want to respect their time.

Chapter 16

HOW TO SET EXPECTATIONS AND CREATE ANTICIPATION BEFORE THE CALL

Once you've filtered out the people who aren't a good fit for your offer, you want to set the right expectations with those who you believe are your ideal target clients.

Specifically, you want to:

1. Remind them of the call (people get busy and forget)

2. Let them know what to expect on the call and to remove any up-front sales pressure

3. Give them an idea of what it's like to work with you with a case study to create anticipation

STAGE 3 ENGAGE

The best way to explain how to engage a potential client before a strategy session or enrollment call is to show you an example of an email I've sent to help potential new clients get ready for a strategy session.

Subject: *Important: Your Strategy Session with Joe Stolte*

Hey Sarah,

Before we hop on for your strategy session, I wanted to send over a couple of things to make sure you're set up for success.

First off, I want you to know that the primary objective of this call is to help you get clarity and focus on how you can get more (and better) clients for your business.

Second, if we both agree we're fit to work together you'll have the opportunity to enroll in our accelerator program. If not, no big deal.

Third, if you're interested in what it's like to work with me and our team you can check out this short video from a past client who got amazing results after completing our accelerator program (click here to see the video).

Lastly, to get the most out of the call I suggest you bring a journal and something to write with. Make sure you're in a place where you can focus (not driving or multitasking).

This call is going to require your full attention for it to be valuable.

If you have any questions in advance of the call you can respond to this email. I'm personally reviewing each message (this isn't an autoresponder).

Be great today,
Joe Stolte

If you don't have a case study, don't worry about it. Revise the above email to fit your situation and personally send it from your business email to your prospective clients before the call.

If you have a sales team, have the person who is conducting the strategy session send this email or a similar message to the prospect directly.

You shouldn't be sending hundreds of these emails if you are doing a good job qualifying people and filtering them out, so you can invest some of the time you're saving by personally engaging people who have made it to this part of your process.

If you absolutely want to get fancy, you can use a 3rd party software to automate sending this email (most scheduling services like Calendly or Schedule Once allow you to send confirmation emails) but I highly recommend you or your team personally respond to any replies you get from prospective clients who respond to this email.

The personal touch differentiates you from the competition and shows that you are meeting them on a human level. It also sets the stage for the next step in the Convert phase which is where all the magic happens.

Chapter 17

THE PERFECT SALES CALL SCRIPT THAT ENROLLS YOUR DREAM CLIENTS IN ONE CALL

This is the most important part of your sales process. It might even be the most important part of your business. If you can learn how to powerfully enroll your ideal target clients into your programs and services, you will never have to worry about money again in your life. You'll be able to get more (and better) clients for the rest of your business career.

Let me start by saying when I got started doing enrollment calls for my consulting practice, I was terrible. I had no idea what to say on the call. I would jump right into telling them about my offer, I got caught up answering all of their questions instead of controlling the conversation and gathering enough data to find out if I could even help them in the first place.

That is until I learned what I now call the "perfect sales call" script. By "learn" I mean I read an endless number of sales books, paid to study under other coaches and elite sales professionals, and then battle-tested what I learned on thousands of sales calls. This script has the benefit of me and my team doing thousands of calls, making every possible mistake, correcting those mistakes, and distilling it all into a powerful framework that works extremely well for selling professional services.

STAGE 4 - ENROLL

The script is designed to powerfully guide your target prospect through a process to diagnose if they're fit, invite them into a new level of clarity on their situation, and then intelligently position your services as the bridge between where they are and where they want to be.

No high-pressure sales tactics, no endless objection handling, no bait and switch, no hollow promises, no used car salesman tactics. Just an open and honest dialogue between you and your ideal target client and a powerful invitation for them to invest in their business to level up.

The 'Perfect Sales Call' script has 7 core sections:

1. Open & Set The Stage

2. Discover The Motive

3. Diagnose The Situation & Pain

4. Define Desired Results

5. Summarize The Gap

6. Get Permission & Pitch

7. Collect Payment

BEST PRACTICES & RULES OF THE ROAD

Before we dive into the actual script, there are a few important best practices and rules you should understand:

- **Diagnose before you prescribe**: Your job is to diagnose the situation before you prescribe your services as the cure. If you walked into the doctor's office and your doctor handed you a prescription without asking you any questions, you would immediately lose trust. In fact, in the medical world prescription without diagnosis is called malpractice. The same principle applies to enrollment calls with your ideal target client.

- **Control the call**: This person is coming to you because they don't know how to do something on their own and they are looking for help. The best way you can help them is by taking control of the conversation and asking them questions, not the other way around. Your job is not to answer questions about your services until you have determined you can actually help. You don't go into your doctor and start asking questions about a medication they haven't prescribed to you yet. The same is true on an enrollment call with your prospective client.

- **Listen closely and take notes**: You will be asking a lot of questions and it's really important that you take notes. One of the most powerful things you can do as a 'helper of humans' is to allow people to feel heard. By listening, taking notes, and paraphrasing back what you heard in the words of the other person, you make them feel heard, welcomed, and understood. This is the exact environment you want to foster on your enrollment calls.

- **Get comfortable with silence**: Ask one question at a time and wait for their response. Get comfortable with silence and don't talk after you've asked a question. If they take 30 seconds to respond, you can ask them if they heard the last question and if they want you to repeat it but don't move on to the next question or bail them out from an uncomfortable silence. Silence is your best friend on a sales call.

- **Don't solve their pain**: You're going to invite them into clarity on their current situation and you will hear them describe what they are struggling with. Being the purpose-driven, heart-centered helper of humans that you are, your first instinct will be to jump in and problem solve and provide solutions and suggestions. Don't do it. You are not helping them by doing this, you are robbing them of an opportunity to find clarity. Your job is not to solve their problems on this call, it's to help them get clear and enroll them as a client if they're a fit.

At the end of the day, client enrollment is heart-to-heart, human-to-human communication.

If you're doing your part, the person on the other end of the conversation is going to walk away from the call with incredible clarity on their situation in a way they haven't been able to get on their own. That is how you can demonstrate you can help them, by actually helping them.

The 'Perfect Sales Call' script is simply a guide to help you sharpen your thinking and to avoid some of the common mistakes we all make when enrolling clients. As with everything in this book, you'll have to take the script and customize it to fit your business, your offer, and your ideal target client.

PRE-CALL PREPARATION

1. Before the enrollment call, take 3-5 minutes to get settled in and prepared. Specifically, make sure you're in a quiet room without distractions.

2. If possible, use headphones so you can keep your hands free to take notes.

3. Make sure you record the call.

4. Have the application answers in front of you along with something to take notes with

5. Take some deep breaths, release all of your attachment to getting a sale, and shift that energy into doing world-class diagnoses.

OPEN & SET THE STAGE

This step has three parts:

1. Open up with some small talk, be human, be relatable, take some time to get yourself and the other person comfortable.

2. Take control and set the agenda so the target client knows exactly what to expect

3. Get their buy-in on the agenda and call format with a clear 'yes' before proceeding.

What I've found works best after the small talk is to say something like this:

"Okay Sarah let's get started. We only have 40 minutes and I want to make sure that by the end of our discussion today that you've gotten a ton of clarity and value from the call…"

This is where you start taking control of the call and guiding the process so you can add real value. Next, you want to make it clear what to expect on the call.

"Let me briefly explain how today's call will go. I'm going to be asking some questions about your business and your application, which I have printed out in front of me, so I can better understand your situation and needs.

Then if it sounds like I can help and if it sounds like we're both a good fit for each other, I'll explain what we have and how it works.

After this, if it makes sense, you can make a decision whether or not you want to move forward or not. Does this sound good?"

You want to wait for a verbal yes here. This brings the other person into the conversation and makes them feel like they are part of the process and helps put them at ease around what to expect.

A lot of people get anxiety or a sense of worry when they're on a sales call. They fear they are going to be taken advantage of. We want to immediately clear the air here and transparently lay out what they can expect. So wait for the yes before you move on to the questions.

UNDERSTAND THE MOTIVE

In this section you want to understand a few key pieces of information:

1. Why are they on the call?

2. Why do they want to make a change now?

To start this section, you simply want to transition from the last section into the first question. Remember they just said "yes" to your last question.

"Great, so let's get started. So, Sarah, tell me what motivated you to get on this call with me today?

This question is designed to understand their motive. When you ask this question, you probably aren't going to get the answer you want.

You're more likely to hear them explain how they found you (webinar, referral, social post, direct outreach, etc). If you get this answer or any other answer that doesn't directly answer your question, then you want to agree, transition, and rephrase the question.

"Ok perfect, I'm glad the post I made on Facebook resonated enough for you to book this call today. So, tell me, why now? What's going on that made you decide to take action on this right now?"

After you rephrase the question, be quiet and let them talk. Silence is golden. You're inviting them into clarity, to be vulnerable about what they are experiencing that got them to get on the call with you. You want to listen carefully and take notes as to what pain or problems they are experiencing to get on the call with you. Do not move on to the next section of the script if you do not get a great answer here. Keep politely asking the question until they open up.

You'll probably need to dig in deeper to move beyond a surface-level response. My favorite follow-up question to do this is "tell me more".

"Interesting, can you tell me more about that?"

Once you get a clear reason for 'why now' you want to repeat what they said to you, so they feel heard. It's really important for the other person to feel heard if you want them to open up further as you diagnose. So, you might say something like:

"So, what I'm hearing Sarah is that your coaching business was generating around $60,000 per month before COVID hit, some of your clients had to stop working with you because of money issues, and now you'd like to get back on track and scale your business to $100,000 or more per month, is that right?"

Again, wait for them to say yes or to correct you if you mis-stated something. Then transition to the next section.

DIAGNOSE SITUATION & PAIN

In this section, you want to dig into their current situation and find out exactly what problems or pain they're experiencing so you can determine if you can actually help them. You also want to summarize what you've heard to ensure alignment and to make the other person feels understood. Specifically, you want to:

1. **Understand the Current Situation**. If you're working with businesses, you want to understand what's going on

in their business. If you're working with clients on weight loss, you may want to understand their diet, exercise, and lifestyle.

2. **Understand the Pain Points**. Get clear on the exact pain they are experiencing and understand what they need the most help with.

3. **Summarize**: Paraphrase back in the client's words what you heard about their situation and pain points. The key is to make sure you're aligned and that they feel understood.

Picking up where you left off in the last section, you want to jump right into a handful of questions to understand their current situation. You'll need to customize this section to fit your offer and your ideal target client. For our business, the questions I typically ask are as follows:

SITUATION QUESTIONS:

*   *Tell me about your offer. What are you selling?*

*   *Let's talk about your pricing, how much are you charging?*

*   *Ok great, who is your ideal target client?*

*   *What problems are your ideal target clients facing that motivate them to buy from you?*

*   *Walk me through the process you take to transform people from strangers to paying clients?*

*   *Why do clients buy from you? What's the promise you're making to the marketplace?*

PAIN QUESTIONS:

- *How much does it cost you to acquire a new client?*

- *Do you feel like your existing marketing and sales process allows you to consistently generate new clients like a machine?*

- *What do you think (insert pain point) is costing you each month?*

- *How else is (insert pain point) showing up for you?*

- *Exactly how much revenue did you generate last month?*

By asking these questions, listening, and taking good notes, you'll begin to understand exactly what's going on with them, the pain they are experiencing, and exactly how it's showing up in their business and personal life.

The next step is to recap what you heard so you are aligned, and they feel understood.

"Ok great Sarah, so what I'm hearing is that you're currently generating $40,000 a month after losing some clients to COVID, you sell a mixture of 1:1, group coaching, and done for you consulting services and you don't have an established pricing structure at the moment.

You're currently getting most of your new business from referrals, but you don't have a consistent system for generating qualified leads each month and it's literally keeping you up at night and causing you to lose sleep and creating more stress at home with your family, and ultimately you think not having a consistent system for getting more and better clients is costing you at least $25,000 in additional revenue per month, does that sound right?"

At this point, you want to wait for the person to say "yes" before you proceed. Once you have a verbal agreement from the other person and they feel understood, move on to the next section.

DEFINE DESIRED RESULTS

Next, we want to transition right into understanding their big picture vision. The objection of this section is to help your prospective client get clear on exactly what they want to accomplish. This is incredibly valuable as most people don't know what they really want.

They have a vague intention or aspiration that they haven't spent time clarifying. In this section, you're going to use a powerful question to help them get a new level of clarity that they haven't been able to get on their own prior to this call. This is one of the core ways this process actually helps your ideal target client and builds goodwill, whether they enroll with you or not.

This also helps you understand if their desired results are aligned with what your services can actually deliver. Remember the goal is to find out if you can actually help before you start pitching them on your solution.

Picking up after getting a 'yes' from the person, you want to say:

"Perfect, so Sarah let's go 12 months into the future, tell me what the perfect situation looks like for you. That is, what has to have happened for you in the next 12 months for you to look back on this conversation and feel like you made real progress in your business?"

This is a powerful question to really help the other person get clarity on what they want. It's important to pause after you ask this question and let them answer you fully. Don't cut them off, just let them talk, and take notes.

Side note: This is a derivation of the classic "Dan Sullivan Question", which was created by Dan Sullivan the creator of the Strategic Coach program.

When they have answered fully, paraphrase back to them and go deeper to understand why these matters to them.

"That's great, Sarah let me playback what I'm hearing to you. What I'm hearing is that in 12 months you want to scale your business to $1M in revenue per year and you want to be able to sleep better at night knowing that your marketing is systematized and bringing in high quality leads without you doing all the work yourself, you also want to start working with better clients who aren't as impacted by the pandemic and a dip in the economy, is that right?"

Again, wait for a verbal yes before you proceed. Once you have the yes, it's time to go deeper.

"Great, so let me ask you, why $1M in revenue per year, what is it about $1M that stands out for you?"

The goal here is to dive deeper into the core motivation. We're going to keep following up with questions until we go 3 layers deep to find their core motivations.

"Ok, so you want to hit $1M annual revenue so you can afford to buy a bigger home, have a second child, and be able to hire help to homeschool your son. So, I'm interested, why is it important for you to homeschool your son?"

Here again, you are digging a layer deeper to find the real reasons why this vision matters.

"Thanks for sharing that with me Sarah, so it sounds like with all the chaos being caused by the pandemic you're concerned about the safety of your son in a public school and you really believe that giving your son access to a private home school environment is a more effective way for him to learn and grow up to be a responsible adult."

Notice how we went from "I want to make $1M in revenue" to "I want to homeschool my son because I'm concerned about his safety in the public school system". The former is a surface-level desire that nearly every small business owner has, and the latter is an incredibly clear, personal, and emotional reason for wanting to grow their business.

You want to dig and keep asking until you get beyond the surface desires and understand the real emotional reasons for why they want to grow their business or achieve the surface level result before you go on to the next section.

SUMMARIZE THE GAP

At this point, you've covered a lot of ground in the conversation and it's time to summarize the primary gap between where they are and where they want to be and find out what is really holding them back.

IN THIS SECTION YOU HAVE THREE OBJECTIVES:

1. Accurately summarize the gap between where they are and where they want to be

2. Diagnose what the prospect believes is holding them back from closing the gap

3. Determine if and when the prospective client wants to fix their situation

This is a good place in the conversation to make sure the prospective client clearly sees and acknowledges the gap between their current situation and their desired results. As the outside observer, this gap may be obvious to you but when you paraphrase it back to the client it's often a very clarifying experience.

It's also important to understand what the client believes is holding them back from closing the gap. If they're not taking any ownership for what's within their control to close the gap, there's a very good chance they won't believe you can help them. If, on the other hand, they know they're playing a role in not getting what they want, that's a good sign that they're coachable and you can actually help them.

Lastly, you want to establish if this person actually wants to fix their situation and when they want to get started fixing it. If they're ok with staying where they are in their current situation or if they believe this is something they can put off for a while, they aren't ready to enroll with you. If so, that's completely fine, you can end the call politely and explain why you think they won't be a fit to work with you.

"We've covered a lot of ground in the call so far, thanks for sharing with me. So, it sounds like you're currently generating about $40,000 in your business per month and in the next 12 months you'd like to be generating more like $84,000 per month so you can cross the $1M mark in total revenue, get a bigger home, have a second child, and homeschool your son to help make sure he's safe and getting the right education" ...

Tell me, Sarah, what's stopping you from achieving this vision on your own?"

At this point in the conversation, you want to pause, listen, and take notes. Specifically, you want to listen for the following reasons for the gap:

1. **Knowledge**: They don't know how to do it on their own

2. **Speed**: They have an idea of how to do it but want to go faster with the right help

3. **Guidance**: They want to follow a proven process from an expert

If their answer fits in one of the above reasons, this is a good sign they are open to getting help, and that you can help them.

Next, you need to find out when they want to get help.

"Okay great, tell me, Sarah, when do you want to get started in closing this gap so you can realize your vision?

Most of the time at this point in the conversation, they are going to say something like "ASAP" or "I want to get started immediately" which is exactly what we want to hear before we move on to the next section.

GET PERMISSION & PITCH

Now you're closing in on the final few sections of the enrollment conversation where you'll get permission from the prospective client to tell them about what you offer and then present your offer.

Getting permission is critical before you launch into selling your services. When selling or enrolling clients, most service providers make the mistake of pitching too early in the conversation.

By pitching too early so you violate Mental Model #1: Make everything about the client and what they want. If the prospective client doesn't want to hear about your offer, don't try to sell them on it.

Here's how you ask for permission to talk about your offer.

"Ok Sarah based on what you've told me I definitely think I can help you achieve your vision. Would you like me to tell you what I do?"

The good news is that if you've done a great job on the call and this is in fact your ideal target client, 95% of the time the prospect will be excited to hear about your offer. Assuming that's the case, now you want to move into your pitch to present your offer.

"Great, well my area of expertise is I help coaches and consultants consistently get more and better clients coming to them so they can scale their business to $100,000 a month or more in revenue."

I typically work with people who need to simplify their business and who need to get more and better clients so they can generate consistent and dependable levels of revenue so they can take care of their family and contribute to their community."

What I've done here is presented my positioning statement with some additional context that specifically fits the situation of the person I'm talking to. This minor customization will be easy to do on the fly if you've been listening, taking notes, and paraphrasing back during the conversation.

The key is to state your area of expertise in a way that speaks directly to the prospective client based on what they have told you during the conversation. If you do this correctly, it will sound perfect to them and they will naturally want to learn more about how it works.

At this point, you want to pause and wait for the prospective client to ask for more detail. This pause and moment of silence will feel awkward when you first try it. Don't talk. Put the phone on mute if you have to.

You want the prospective client to ask you for more information before you start going into the details of your offer. This is important to ensure alignment, interest, and desire before you launch into the rest of the pitch.

Once you hear them say something along the lines of "how does it work?" or "ok, great, how do we get started", you want to tell them about your offer.

In this section, you will have to customize this for your offer. The important things to focus on when summarizing your offer are:

1. Keep it brief (no more than 2 minutes)

2. Don't talk about the 'how' talk about the 'what' (you aren't here to solve the problem on the call)

3. Don't talk about price yet (that comes next)

Here's an example of how to present your offer:

"Well, we take clients through a 90 day 1:1 coaching program that covers the four key phases of getting more (and better) clients. The program breaks each phase down step by step and teaches you exactly how to simplify and package your offer, how to craft magnetic messaging that gets clients consistently coming to you who are excited to work with you and want to pay you upfront. We then help you find the perfect channels to deploy your messaging into and help you enroll clients quickly. If you're the kind of person who takes fast action, you should have an increase in leads and new sales within the first 21 days of joining the program."

Now it's time for another pause. Remember silence is golden. You want to give the person time to process and you want them to lean in for more information vs. you jumping right into giving the price. If they're interested, they will ask for more information. The goal is to have them tell you what information they need to move forward, not for you to drop endless amounts of data on them and risk confusing them or overwhelming them.

At this point in the conversation, you'll need to answer each question and address each possible concern. You don't want to talk about price until they ask.

If they don't ask a question, don't respond. Let them talk, let them process, and let them consider the conversation. You are putting them in control. This is not what most people are used to in an enrollment conversation. They are used to being hard sold and they are used to being 'handled' or 'closed' in a conventional way. That's not the game we are playing with them. We simply want to give them space to consider what you've said and to ask questions.

Once you've answered all their questions, they will naturally ask you how much your program or services cost. At this point, it's finally acceptable for you to talk about price.

"The investment in the accelerator program is $10,000"

State your price, then shut up. Mute if you have to. Don't talk. Let them make the decision and talk themselves into doing it or decide if it's not right for them.

At this point, 1 of 4 things is going to happen, they will:

1. Ask more questions, to which you should provide answers

2. Tell you it's too expensive or they don't have the money

3. Tell you they need time to think about it

4. Indicate that they want to move forward by asking about the next steps or how to get started

In the case of #2, they tell you it's too expensive or they can't afford it, here is what I like to say:

"I hear you, it's a serious investment. Let me ask you an honest question, do you really want this and believe it can help you achieve your vision? Sometimes saying it's too expensive is just a polite way of saying you don't really want it."

What I'm trying to accomplish here is to understand if they aren't sold or if it's truly a money issue. If they aren't sold, I will ask them what part of the offer they think won't help them get results so I can make sure they understand it all clearly.

Most of the time they are sold and it's a money issue. If that's the case I will tell them about our payment plan and ask if they want to get started. If not, I politely end the call.

In the case of #3, they need time to think about it, I want to make sure they really want it before we talk about the next steps and follow-up. If they don't want to enroll, I don't want to waste their time or mine.

"I totally understand wanting time to think about it, let me ask you a candid question though, do you really want to enroll or is this a money situation?"

If it's a money situation, I will revert to the talk-track laid out in the previous example.

If they really want to enroll and truly want time to think about it, I'll simply ask them how much time they need exactly, and then set up a follow-up call for me to check-in on them.

In the case of #4 where they're ready to move forward, you want to collect their payment over the phone or send them a link to make the payment while you are on the phone. You can simply say something like the following:

"To get started, I can take your credit card details over the phone now, and then we can discuss the next steps to get you started right away. I'm ready to take your card now if you have it handy?"

Or

"I'm going to email you a link to a secure payment site where you can make your payment so we can get started quickly. I'll stand by while

you open the email and make the payment. Then I'll tell you exactly what to do so you can get started in the program immediately."

At this point, you want to take their card or wait for them to make the payment, then tell them exactly what they need to do in order to get started in the program or how to move forward with your services.

Congratulations, you just powerfully and ethically enrolled a new client.

Chapter 18

THE NEW SCIENCE OF FINDING YOUR MESSAGE MARKET FIT PART II

Before we conclude the Convert Phase of the Tractionology framework, it's important to remember that our number one objective of the enrollment conversation isn't to actually get the sale or to enroll the client (yet).

At this stage, the number one goal is to validate your minimum viable magnetic messaging to ensure you have 'message market fit' so you can consistently get more (and better) clients.

You're going to want to do about 30 enrollment calls to get a sense of how well your minimum viable magnetic messaging is landing and to look for themes and trends based on what your ideal target clients are saying on your sales calls.

Specifically, you want to keep track of the following 8 questions:

1. What words or phrases do you hear your ideal target clients using over and over again on sales calls?

2. What stories and analogies do your ideal target clients seem to tell over and over again across multiple calls?

3. What are the core challenges and problems you are hearing during the 'Diagnose Situation & Pain' portion of your sales calls? Do your positioning statement and minimum viable magnetic messaging speak to these pain points?

4. What is the core desire your ideal target clients are sharing with you during the 'Define Desired Results' section of the sales call? Do your positioning statement and minimum viable magnetic messaging speak to these pain points?

5. How are your ideal target clients responding to your positioning statement in the 'permission and pitch' section of the sales call?

6. What words and phrases seem to resonate <u>most</u> with your ideal target clients?

7. What words and phrases seem to resonate the <u>least</u> with your ideal target clients?

8. What changes do you need to make to your minimum viable messaging based on trends you are hearing across multiple calls?

Tracking the answers to these questions is how you're going to continue to validate and optimize your message making it extremely magnetic to your ideal target clients. You will know exactly what to say to grab their attention and to make your messaging incredibly effective.

We've covered a ton of ground in this section. By now you should understand the 4 stages of the Convert Phase, you should know exactly what to say to lead a powerful sales call that converts more (and better) prospects into paying clients, and you should know how to use the sales call to validate and optimize your minimum viable magnetic messaging.

Now it's time to move on to the final phase of the Tractionology framework where you're going to learn the secret my mentor taught me to transform clients into raving fans that send you referrals over and over again.

SECTION V

HOW TO TRANSFORM YOUR CLIENTS INTO RAVING FANS WHO SEND YOU AN ARMY OF HIGH-QUALITY REFERRALS

Chapter 19

THE REAL SECRET TO SELLING PROFESSIONAL SERVICES

"Sit up straight, speak up, look me in the eyes, and start over." I was 3 minutes into my first "partner review" in my second week as a management consultant and things were not going well.

I swallowed hard, did as instructed, and started again from the top of my presentation. Two minutes later I heard him say "there are no page numbers on these slides, the bullets aren't evenly formatted, and I can't read any of this. Go find someone else to help you review and practice before you submit something like this to me again". The meeting was over.

As the first-ever 'summer intern' for a new management consulting firm, I was getting my first lesson in professionalism and client service. The person delivering the hard feedback was the Managing Partner who also happened to be my mentor.

It took several more sessions like this over the next year for me to get the lesson and understand the real 'secret' to selling professional services.

About a year later, I was talking to the same mentor and I asked him what it took to become a partner at the firm. He gave me a smile that told me he knew exactly where this conversation was going and explained that there are two paths to become a partner at any consulting firm.

The common path is to put in seven or eight years and work your way up through promotions and doing well in performance reviews. The less common path, the path that few are willing to talk about, is to sell more consulting engagements to grow the practice revenue. In this second path, you still have to learn and develop as a professional in your career, but the promotions come much faster.

Naturally, my next question was "great, how do I learn how to sell more consulting engagements?" and this is when I learned the real secret to getting more (and better) clients.

This secret has landed me more work as a consultant, coach, and entrepreneur than anything I've explained in this book so far. It's the reason why I will never have to worry about money or finding clients again as long as I'm alive and working.

As we walked across the street and back into the office, he told me the secret in 3 simple words.

Do good work.

I looked at him puzzled for a moment before he continued on.

"The best way for you to sell more professional services is to do good work. It's for you to care about the client and make sure that you're delivering beyond what is expected".

This profound wisdom was almost too simple for my 20-something-year-old brain to understand but the more we talked about it, the more it began to sink in.

By 'doing good work' clients would get results and buy again (and again and again and again). Years later I would learn that clients would not only buy again, but they would also refer other clients who would buy again and again.

Doing good work means caring about your clients. Caring about your clients means doing the small things right, consistently. This is why not having page numbers on my slides and having bullets that weren't consistent mattered so much. It was a proxy indicator of my level of excellence and care for the client.

This strategy worked incredibly well for my mentor and his firm. They grew from about 10 staff members to well over 70, with most of that growth taking place during one of the worst economic recessions in U.S. history. Years later they would go on to sell the firm to the 6th largest professional services company in the world.

It's worth mentioning that the firm didn't run a single paid ad, they didn't have a digital funnel, and they didn't have complicated email autoresponders. What they did was deeply care about their clients and do incredibly good work.

TRANSFORMING YOUR CLIENTS

Once your clients have enrolled, your job of 'selling' is just getting started. Your objective shouldn't just be 'get the sale' but to transform your clients into raving fans who will become repeat purchasers that regularly send you an army of high-quality referrals.

We don't want to leave the client transformation up to chance though, we want to carefully architect the client journey, so your clients are guided to want to buy again and to systematically send you referrals. To facilitate this process, there are 3 key things you need to do:

1. **Deliver Results**: Get the clients the results they are looking for.

2. **Optimize For First Value**: Get clients to the 'ah-ha' moment quickly and efficiently.

3. **Systemize Growth**: Build systematic growth into your fulfillment process.

Chapter 20

THE TRUTH ABOUT DELIVERING CLIENT RESULTS

You can execute every fancy trick in the book, but if you can't help your clients get the actual results they're looking for then you can't transform them into raving fans.

Sadly, this is where so many service providers in the internet marketing world tend to fall down. They put an incredible amount of focus on the magic words to get you to buy, the fancy 'trip-wire' funnel, and the endless 'one-time offers' in the sales process and they neglect to deliver on the core promise of getting you the results you want. They drown you in bonuses and more 'information' at the expense of getting you focused on the shortest and most direct path to getting you results.

If you want to transform your clients, you need to deliver exactly what the client needs to get them from where they are to where they want to be and keep them focused on this path.

TO DELIVER CLIENT RESULTS, FOLLOW THESE 5 PRINCIPLES:

1. **Know Your Ideal Target Client**: You have to know exactly who you serve, what results they want, and then

deliver products and services that help them get what they want. This is why we start the Tractionology framework by helping you identify your ideal target client. The client who you would happily take on, even if you could only get paid after they got the results they wanted.

2. **Only Work With Your Ideal Target Client**: Your client needs to do their part in getting the results, otherwise they won't be transformed. This is why it's critical to filter out and exclude anyone who is not your ideal client. If a prospect doesn't fit your criteria when they fill out the application or jump on a sales call, don't take them as a client.

3. **Construct a Powerful Core Offer**: Your offer should have the right steps to get someone from where they are to where they want to be. If it's missing something that is on the critical path to helping your client get results, add it in. If you have things in your offer that don't directly contribute to getting your client the results they want, take them out. Less is more, simple is better.

4. **Stay Focused**: Don't make the mistake of thinking your offer is like a grocery bag that needs to be filled to the top with all the different stuff you can offer. Having a lot of fancy bonuses may induce the client to enroll but it also overwhelms them and drowns them in useless information. What is the minimum effective dose of services, products, information, and support your client needs to get what they want? Start there and perfect that before you get fancy.

5. **Test & Iterate**: Pay careful attention to why some clients get incredible results and why others seem to struggle

despite your best efforts. Look for the patterns, themes, and trends in your fulfillment that directly contribute to helping your clients transform and get results. Also look for the patterns, themes, and trends that are taking away from clients getting results. Craft the client journey to maximize their chances of getting results.

Through testing and iterating in our company, we learned that some clients are so overwhelmed and burned out when they reach us that before they have the capacity to implement what we teach, they need to optimize how they spend their time so they can eliminate distractions.

To help clients eliminate distractions, we created a short and effective training module around how to free up their time and create space to do the work necessary to get more (and better) clients.

We don't advertise this training and it isn't a 'bonus', it's an essential step to help certain clients get results and we discovered it by following the principles listed above.

Helping your clients get the results you promise is essential to transforming them into raving fans but it's not enough. You want to architect your client's experience so that they get value quickly, build early momentum, and ultimately buy more from you and send you a ton of referrals. To do this, you need to track and optimize a key metric called 'time to first value.'

Chapter 21

SILICON VALLEY SECRETS FOR SCALING YOUR SERVICE-BASED BUSINESS

One of the big take-aways I learned from scaling two technology software companies to over 70 employees and thousands of customers is the concept of 'first value' or the ah-ha moment.

"First value", also known as the 'ah-ha moment', in a software product is the moment of sudden insight or discovery when a new customer realizes the value of the product and why they need it.

The faster you can get a new customer or client to have their 'ah-ha' moment, the more likely they are to come back and use your product or service over and over again.

This concept is exactly how Facebook scaled to its first 1 billion users. Their 'first value' or ah-ha moment was getting you to connect with 7 friends in 10 days.

Facebook knew that if they could get you to connect with 7 friends in 10 days, you would be hooked for life. This was their core focus and strategy with the product in the early days and has been credited as one of the core drivers of their wild success.

For slack, the popular messaging application, the first value metric is 2,000 messages. Slack knows that if they can get a team onboarded and using the product past the point of sending 2,000 messages, then the team will have their ah-ha moment and be hooked for life.

For Twitter, it was getting a new user to follow 30 other users.

For a former client of mine that helps coaching companies hire their first sales closer, the first value moment was getting his clients to the point where they closed their first high-ticket client without the owner being on the phone all day.

If you can identify your client's first value moment and optimize their experience with you to achieve this moment as efficiently as possible, you're going to transform a lot more clients and help them get results faster.

To optimize for 'first value', you need to execute the following 4 steps:

1. **Define your point of first value**. Examine the history of your most successful clients. Is there a clear moment or experience they had after working with you where they finally "got it" and the lights when on? Is there an early moment in their journey of transformation where they got a quick win that created momentum? If you have a team, talk to them and brainstorm what this moment might be. Talk to past clients who have had extraordinary results and ask them if there was a core moment that stands out for them. Even if you don't get this moment completely right, form a hypothesis and then test it on future clients to see if it holds true. Get feedback from clients and iterate.

2. **Optimize your client onboarding experience**: Carefully script the exact steps your clients need to be taking so they are focused on the shortest and most direct path to experiencing their time to first value or ah-ha moment with you. If there are unnecessary steps, eliminate them or push them to after the point of first value. Make sure every person on your team knows the critical path for getting clients to the ah-ha moment and regularly discuss with your team ways to streamline the client journey even further.

3. **Surface client milestones that go unnoticed**: You want to script and celebrate client milestones early and often to create the feeling of momentum. If the client has taken fast action, celebrate that with them, and acknowledge them for their achievement. I've had the opportunity to serve as a coach in an organization called 2X which was founded by my friend Austin Netzley. One of the things Austin has all his coaches do is to celebrate wins with the client every single week. It's the first agenda item on every single coaching call. This builds momentum for the client and keeps them engaged on their pathway to experiencing first value.

4. **Track and review 'time to first value'**: We recommend our clients start tracking a simple metric called "time to first value" which is the number of days it takes your ideal client to get to their first value moment. The clock starts ticking the moment they enroll with you and ends when they get to their ah-ha or first value moment. This metric should be one of the most important numbers you and your team pay attention to in order to measure the health and success of your client fulfillment. In my team, this is the very first key performance indicator (KPI) we look at

in team meetings to understand how well we are doing with transforming our clients.

By helping your clients get results and by optimizing your client onboarding experience for first value, you're well on your way to transforming your clients into raving fans but your work is not done yet.

If you really want to transform 100% of your clients into raving fans that send you an army of referrals, you need to systemize growth into your service experience.

Chapter 22

BUILD YOUR SYSTEMATIC REFERRAL MACHINE IN 5 SIMPLE STEPS

The next step you want to take in the 'Transform Phase' is to systematize growth into your service experience. You don't want to leave growth up to chance.

If you've helped your clients get results and architected a client experience that gets them to their 'first value' moment quickly, you will have generated a lot of goodwill with your client. The next step is to leverage this goodwill to get more (and better) clients.

Another big takeaway I learned from the world of technology and scaling software applications is the growth concept called "k-factor". In the software world, your k-factor is a fancy term for the number of new users that each new user brings to your product.

When designing a software product, a really good product manager will work to ensure that the product experience induces its users to refer other users to join the platform as well. That way every new user ends up bringing one or more new users along with them. This dramatically lowers customer acquisition costs and can seriously skyrocket growth.

Chances are you've experienced this when onboarding to Facebook, LinkedIn, Twitter, or any major social networks or popular consumer applications. As soon as you join the app, one of the first screens you are shown encourages you to invite other people in your phone's contacts to join the platform with you. This is systemized growth built into the product experience.

In the earliest days of the internet, hotmail.com used an incredibly simple tactic to scale their users from several thousand to several million.

They added the words "PS: I Love you. Get your free email at Hotmail" to the bottom of every email with a URL that linked to their registration page. It took Hotmail about six months to get their first million users but only 5 weeks to get their second million users, largely because they systemized growth into the product experience.

While I was growing Lottery.com, a mobile application that allows users to purchase lottery tickets from their mobile phone, one of the things we added to the onboarding experience was the ability for every new user to invite a friend to the application and by doing so they could unlock a free lottery ticket for them and the person they invited. This one tricked helped us turn one new user into 2 or 3 without any additional marketing spend.

The great news for all of us service-based business owners is that we don't need to sell software to systemize growth into our businesses. We simply need to be intentional around how we craft the client experience to replicate some of the same growth tactics our friends in the software world have successfully used.

For example, my business coach, Bryan Franklin, is world-class at systemizing growth into his coaching practice. Bryan had the opportunity to serve as the coach for several technology CEOs, including Reid Hoffman of LinkedIn, as they scaled from "millions to billions" in growth, giving Bryan a front-row seat to seeing exactly how these organizations scaled.

When I onboarded with Bryan, one of the first things he did was to conduct a "360 review" with my peers and direct reports to gather feedback on my performance. This involved him going around and interviewing each of these people for 30 minutes to gather feedback on my behalf.

This was a valuable part of my transformation but also a clever way for Bryan to announce to others in my company who he was and what he did. This immediately got other employees interested in coaching and asking if they could also receive coaching from Bryan. This is one of several brilliant ways Bryan systemized growth into his coaching practice and it's what has allowed him to continuously add new clients without any kind of paid advertising.

I recently worked with a client of mine who does mindset coaching for medium-sized businesses to replicate Bryan's growth strategy. Within 20 days he had doubled his new qualified leads by adding a modified 360 review process to his client onboarding experience. One simple change with no ads and no cold outreach.

To get started systemizing growth in your client-based business, do these 5 things:

1. **Be referable**: Care about your clients and do good work so you're starting from a powerful foundation of existing goodwill. This makes it a lot easier for others to refer you because they can reference the results you've helped them achieve.

2. **Make it easy**: Make it as simple as possible for your clients to refer you to other people who they think would be a good fit. For example, use forwardable email templates and reusable collateral that your clients can start with to make their life easier when referring you.

3. **Use incentives**: Incentivize both the referring client and the person they are referring to you. In the software world, we call this a 'double viral loop' because both parties get something to engage with you.

4. **Target the right people**: Make it clear who your ideal target client is so you can maximize the chances that your existing clients will send qualified leads your way. This is why it's critical to have your positioning statement dialed in as part of the "Package Phase".

5. **Use The 100% Rule**: My friend Pat Bennett who is head of marketing and sales at 2X likes to ask "100% of clients for referrals 100% of the time". I've taken this a step further in my own coaching practice systemizing a request for referrals at 3 key times in the client's experience: Once when you join, once when you pass your time of first value, and once when we wrap up an engagement together.

By delivering results, optimizing for 'first value', and systemizing growth into your service-based business you're not only transforming your clients, but you're also building a powerful growth engine into your business to consistently get more (and better) clients.

When implemented properly, the tools we've unpacked in the "Transform Phase" can easily double your business over time while saving you a ton of time and stress.

Chapter 23

NOBODY CARES WHAT YOU'RE CAPABLE OF. THEY CARE WHAT YOU HAVE THE COURAGE TO DO.

For years I knew I was capable of more. I knew I had something to offer the world. I knew that I had unique gifts and that I wanted to be of service. I knew that my gifts, fully expressed, would come in the form of me leading a team and using business as a vehicle to change lives.

The hard truth I had to learn was that no one cares about my capabilities or my natural gifts or my unique talents. They only care about what I have the courage to actually do. They only care if I express my capabilities to make them real in the physical world.

The same is true for you.

If you have the courage to apply what you have learned in this book, you will grow your business. You will get more (and better) clients. You will have an impact and you will help people truly transform their lives.

But this is going to require you to have enough courage to take action. This is going to require you to take the knowledge in this book, combine it with your natural talents and gifts, and deploy them into the world through consistent action.

Let me ask you an honest question.

Can you imagine what would happen to you if you applied what you've learned in this book to your business and to your life?

What would that actually be like? Think about it for a minute.

Imagine if you…

- Never had to fumble for the right words to describe what you do and who you serve and instead had a razor-sharp message that made the hair on the back of your ideal client's neck stand up.

- Didn't have to juggle multiple offers and could focus on your best core offer that makes your life easy and transforms your ideal target clients.

- Never had to guess what to say or what content to create to get your ideal target clients coming to you vs. you going out and chasing them.

- Deployed 2 or 3 of the attraction channels in this book and could generate a consistent, predictable flow of qualified leads so that your calendar is full of strategy sessions.

- Stopped hearing the words "I need to think about it" and instead had clients asking you "when can we get started?" or "How soon can we get started?"

- Could sleep at night knowing you've mastered the ability to attract, convert, and transform clients at will.

In this book, I've given you the exact roadmap you need to get more (and better) clients. I've shown you the pathway, now all you need to do is to make a powerful decision around what comes next.

In fact, I would say you have 3 options to choose from at this point:

Option 1 - Do nothing. You can go back to what you're doing and discard what you've learned in this book. While I fully respect your right to make this decision, I truly hope this isn't what you ultimately choose to do. What we've covered here in this book is not theoretical, it's validated and proven as is evidenced by my own results and the results of my clients.

Option 2: Do it on your own. You can take what you've learned in this book and get started right away. My intention in this book was to arm you with actionable guidance you can use right away to get more (and better) clients. If you have the courage to take action and follow the exact instructions in this book, I'm confident you will get incredible results for your business and for your clients.

Option 3: Jump in the time machine. You can dramatically shortcut your learning curve and speed of implementation by having me and my team support you in implementing Tractionology in your business. We'll create a custom roadmap specific to your unique business and show you exactly what to do and when to do it based on your specific situation. If you want the shortest and most direct path to more (and better) clients, we're here to support you.

If you're interested in working with us, simply head over to http://www.tractionologygroup.com/apply where you can schedule a time for a strategy consultation. We'll review your application and if we think you're a fit, we'll hop on a call to see if and how we can be of service.

Regardless of what you choose, I want to thank you for giving me your attention and more importantly your time.

I know your time is irreplaceable and it's the most valuable asset you have to invest. I truly hope what you have discovered in this book will give you a positive return on that investment in the form of you getting more (and better) clients.

I also hope what you've learned in this book allows you to serve the world in a bigger way through your client-based business.

To your success,
Joe Stolte
CEO Tractionology Group